Body composition studies are used in a wide variety of fields including human biology, medicine, sports science, epidemiology and nutrition. They may be used to predict later body composition from childhood measures, bone density loss in athletes under heavy training regimes or in the elderly, to assess obesity in children and adults, and to investigate the likely effects of malnutrition.

Recently, there has been a resurgence of interest in the field of body composition, together with rapid development of a whole new range of assessment techniques.

In this volume, new and old techniques are evaluated, and future developments and aims are discussed. The chapters assume little background knowledge, and will therefore be invaluable to nutritionists, biologists, clinicians and medical physicists alike.

SOCIETY FOR THE STUDY OF HUMAN BIOLOGY SYMPOSIUM SERIES: 36

Body composition techniques in health and disease

PUBLISHED SYMPOSIA OF THE SOCIETY FOR THE STUDY OF HUMAN BIOLOGY

Numbers 1–9 were published by Pergamon Press, Headington Hill Hall, Headington, Oxford OX3 0BY. Numbers 10–24 were published by Taylor & Francis Ltd, 10–14 Macklin Street, London WC2B 5NF. Futher details and prices of back-list numbers are available from the Secretary of the Society for the Study of Human Biology.

Body composition techniques in health and disease

EDITED BY

P. S. W. DAVIES AND T. J. COLE
MRC Dunn Nutritional Laboratory, Cambridge

CAMBRIDGE UNIVERSITY PRESS
Cambridge, New York, Melbourne, Madrid, Cape Town, Singapore, São Paulo

Cambridge University Press
The Edinburgh Building, Cambridge CB2 2RU, UK

Published in the United States of America by Cambridge University Press, New York

www.cambridge.org
Information on this title: www.cambridge.org/9780521461795

First published 1995
This digitally printed first paperback version 2006

A catalogue record for this publication is available from the British Library

Library of Congress Cataloguing in Publication data
Body composition techniques in health and disease / edited by P.S.W. Davies and T.J. Cole.
p. cm. – (Society for the Study of Human Biology symposium series ; 36)
Includes index.
ISBN 0 521 46179 0 (hardback)
1. Body, Human – Composition – Measurement. 2. Anthropometry – Methodology. I.
Davies, P. S. W. II. Cole, T. J. III. Series.
[DNLM: 1. Body Composition – congresses. 2. Anthropometry – methods – congresses.
W1 SO861 v.36 1995 / QU 100 B668 1995]
QP33.5.B63 1995
612 – dc20
DNLM/DLC
for Library of Congress 94-5327 CIP

ISBN-13 978-0-521-46179-5 hardback
ISBN-10 0-521-46179-0 hardback

ISBN-13 978-0-521-03192-9 paperback
ISBN-10 0-521-03192-3 paperback

Contents

Contributors

P. Brambilla
Clinica Pediatrica IV, Ospedale San Raffaele, Via Olgettina 60, I-20132 Milan, Italy

G. Chiumello
Clinica Pediatrica IV, Ospedale San Raffaele, Via Olgettina 60, I-20132 Milan, Italy

C.B. Christ
Department of Kinesiology, University of Illinois, 125 Freer Hall, 906 South Goodwin Avenue, Urbana, IL 61801, USA

P. Duerenberg
Department of Human Nutrition, Wageningen Agricultural University, Bromenweg 2, NL-6703 HD Wageningen, The Netherlands

M. Elia
Dunn Clinical Nutrition Centre, 100 Tennis Court Road, Cambridge CB2 1QL, UK

C.J.K. Henry
School of Biological & Molecular Science, Oxford Brookes University, Gipsy Lane, Headington, Oxford OX3 0BP, UK

S.B. Heymsfield
Obesity Research Centre, Weight Control Unit, 411 West 114th Street, New York, NY 10025, USA

S.A. Jebb
Dunn Clinical Nutrition Centre, 100 Tennis Court Road, Cambridge CB2 1QL, UK

P. Manzoni
Clinica Pediatrica IV, Ospedale San Raffaele, Via Olgetinna 60, I-20132 Milan, Italy

N.G. Norgan
Department of Human Sciences, Loughborough University, Loughborough, Leicestershire LE11 3TU, UK

J. Parizkova
IVth Clinic of Internal Medicine, U Nemocnice 2, Prague 2, 12000 Czech Republic

E.M.E. Poskitt
Dunn Nutrition Group, MRC Laboratories, Fujara, Banjul PO Box 273, The Gambia, West Africa

A. Prentice
MRC Dunn Nutritional Laboratory, Downham's Lane, Milton Road, Cambridge CB4 1XJ, UK

M.F. Rolland-Cachera
ISTNA-CNAM, 2 Rue Conte, F-75141 Paris Cedex 03, France

S.J.S. Ryde
Swansea *In Vivo* Analysis Research Group, Department of Medical Physics and Clinical Engineering, Singleton Hospital, Swansea SA2 8QA, UK

P.S. Shetty
Centre for Human Nutrition, London School of Hygiene and Tropical Medicine, Keppel Street, London WC1E 7HT, UK

P. Simone
Clinica Pediatrica IV, Ospedale San Raffaele, Via Olgetinna 60, I-20132 Milan, Italy

M.H. Slaughter
Department of Kinesiology, University of Illinois, 125 Freer Hall, 906 South Goodwin Avenue, Urbana, IL 61801, USA

M.A. Smith
Centre for Bone and Body Composition Research, Academic Unit of Medical Physics, Leeds General Infirmary, Great George Street, Leeds LS1 3EX, UK

J.F. Sutcliffe
Centre for Bone and Body Composition Research, Academic Unit of Medical Physics, Leeds General Infirmary, Great George Street, Leeds LS1 3EX, UK

Zi-Mian Wang
Obesity Research Centre, Weight Control Unit, 411 West 114th Street, New York, NY 10025, USA

1 *Application of dual-energy X-ray absorptiometry and related techniques to the assessment of bone and body composition*

ANN PRENTICE

Introduction

The development of absorptiometric techniques has revolutionised our ability to measure the mineral content of the human skeleton in living individuals. The first technique to be introduced was single-photon absorptiometry (SPA), capable of the precise measurement of bone mineral in the appendicular skeleton (Cameron *et al.*, 1962; Cameron & Sorenson, 1963; Sorenson & Cameron, 1967). Some years later, dual-photon absorptiometry (DPA) was developed for the measurement of bone in the axial skeleton (Mazess, 1971; Roos & Skoldborn, 1974). Versions of these instruments are capable of scanning the entire body and these provide a measure of whole-body fat and lean masses in addition to total-body bone mineral content. The photon beams required for SPA and DPA are produced by gamma-emitting radioisotopes. In recent years, X-ray technology has replaced the use of gamma-rays and the instruments are referred to as single- or dual-energy X-ray absorptiometers (SXA, DXA).

Absorptiometric techniques are known by a variety of names. For example, bone densitometry refers to any absorptiometric method which measures bone mineral content; X-ray absorptiometry (XRA), quantitative digital radiography (QDR) and X-ray spectrophotometry are synonyms for DXA, and there is a continuing debate as to whether the accepted abbreviation for dual-energy X-ray absorptiometry should be DXA or DEXA.

Currently, the principal application of absorptiometry is for the measurement of bone mineral content in clinical practice (Johnston *et al.*, 1991). This includes screening for osteoporosis, either as a primary condition or secondary to a number of disease states, and long-term monitoring of therapies and drugs known to affect bone. Imaging of prostheses and stones is possible with the latest DXA instruments. In recent years there has been an increasing use of absorptiometry as a research tool.

Examples of research studies that have involved absorptiometry include investigations of the determinants of peak bone mass and bone loss in later life, assessments of risk factors for osteoporosis, and estimations of nutritional requirements for growth and reproduction. The ability of whole-body instruments to estimate total fat and lean tissue masses has also been exploited in studies of obesity, ageing, sports physiology, anorexia, and wasting due to cancer and AIDS.

This review will describe and discuss the principles of single- and dual-energy absorptiometry, the advantages and limitations of the instruments, and some issues concerning the interpretation of bone mineral data.

Methodology

Single-energy absorptiometry

The principle of absorptiometry is based on the exponential attenuation of penetrating photons as they pass through tissue. When a monochromatic beam passes through a region of the body consisting of two tissues, the intensity of the emerging beam is related to the initial intensity as follows:

$$\ln(I/I_0) = -\mu_a m_a - \mu_b m_b \tag{1.1}$$

where I is the final intensity, I_0 the initial intensity, μ_a the mass attenuation coefficient of tissue a, m_a the mass of tissue a, μ_b the mass attenuation coefficient of tissue b, and m_b the mass of tissue b. The mass attenuation coefficients are constants that are characteristic of the tissues at the energy of the photon beam.

In single-energy absorptiometry, as applied to the measurement of bone mineral, the amount of soft tissue in the beam has to be known in order to quantify bone mass. This is achieved by packing the scan region with tissue-equivalent material, by immersing the limb in a water bath, using a tissue-equivalent gel, or wrapping a supple bag of tissue-equivalent material around the limb and using moveable plates to flatten the upper and lower bag surfaces. This ensures a constant overall thickness of the scan region. The baseline intensity is measured as the intensity of the beam after passing through the soft tissue region immediately adjacent to the bone. This allows for differences in soft tissue thickness and composition from one scan to another. The composition of the soft tissue over- or underlying bone has to be assumed to be homogeneous and of identical composition to the adjacent soft tissue. A consequence of these requirements is that single-energy absorptiometry cannot be used for regions of the body where the skeleton is surrounded by soft tissue of variable thickness and composition, such as in the trunk, and, in practice, it is used only for the bones of the arms and legs, particularly the radius, ulna, humerus and femoral shaft.

The mineral content in a transverse cross-section of the bone of interest, 1 cm wide in the axial direction, is obtained by measuring the emerging intensity of the finely collimated photon beam at frequent intervals along the scan path. The final equation becomes:

$$\mathrm{BMC} = [Lp_b/(\mu_b p_b - \mu_s p_s)] \sum \ln(I_0/I) \qquad (1.2)$$

where BMC is the bone mineral content (g/cm), L the data acquisition interval (cm), μ_b the mass attenuation coefficient (cm^2/g) of bone, p_b the density (g/cm^3) of bone, μ_s the mass attenuation coefficient of soft tissue, p_s the density of soft tissue, I_0^+ the baseline intensity after passage through soft tissue, and I the emerging intensity after passage through bone and soft tissue (Sorenson and Cameron, 1967; Wahner *et al.*, 1984; Wahner & Riggs, 1986; Tothill, 1989).

In general, the monochromatic photon beam in SPA is provided by the radioisotopes ^{125}I or ^{241}Am, and detection is with scintillation crystals, usually sodium iodide (NaI). ^{125}I produces gamma-rays of energy 31 and 35 keV, which after passing through a tin filter provide a monochromatic beam of energy 27.4 keV. At this energy there is good contrast between bone and soft tissues while the attenuation characteristics of muscle, water, tendons and blood vessels are similar. However, major changes in fat content within bone and in the adjacent tissues can produce difficulties (Wahner & Riggs, 1986). The half-life of ^{125}I is relatively short (60 days), and such sources require replacement every 5–6 months. ^{241}Am produces a gamma-ray of higher energy, 59.6 keV, which has greater penetrating power but lower contrast between bone and soft tissues and is more prone to problems associated with intra- and extra-osseous fat. The greater penetrating power of ^{241}Am makes this isotope useful for scans of the humerus and femur and its long half-life (438 years) means that frequent source changes are not necessary. Single-energy machines based on X-ray technology are now becoming available.

Dual-energy absorptiometry

The simultaneous use of two photon beams with different energies makes possible the quantification of bone mineral in regions of the body where it is impossible to achieve constant thickness and where soft tissue composition may be variable.

The basic equations for the attenuation of each beam as they are transmitted through two tissues are identical to those in equation (1.1) (Peppler & Mazess, 1981):

$$\text{Energy 1: } \ln(I'/I'_0) = -\mu'_a m_a - \mu'_b m_b \quad (1.3)$$
$$\text{Energy 2: } \ln(I''/I''_0) = -\mu''_a m_a - \mu''_b m_b \quad (1.4)$$

The abbreviations are the same as for equation (1.1) with the superscripts ′ and ″ denoting the values at energy 1 and 2 respectively. The mass attenuation coefficients for the two tissues are constants which depend on the beam energy. Measurement of the initial and emerging intensities of both beams passing through the same volume enables the simultaneous equations (1.3) and (1.4) to be solved and the masses of both tissues to be calculated.

Dual-energy absorptiometry makes use of this principle to determine the bone mineral content in anatomical regions of the axial skeleton (Roos & Skoldborn, 1974; Peppler & Mazess, 1981; Wahner *et al.*, 1984; Wahner & Riggs, 1986; Tothill, 1989). Attenuation at two energies in a region of soft tissue gives fat and lean masses; in regions containing bone it provides bone and soft tissue masses. The composition, and hence the mass attenuation coefficients, of soft tissue in regions containing bone is assumed to be identical to that of adjacent soft tissue in these calculations. The bone mineral, fat and lean masses along one scan are built up by repeated data acquisition along the length of the scan, as in single-energy absorptiometry. Rectilinear scanning over the entire region allows the masses within a defined anatomical region to be determined. Some instruments are dedicated to the measurement of bone mineral in the lumbar spine and proximal femur. Others are capable of scanning the entire body, and provide information about total and regional distributions of bone mineral, fat and lean tissues.

In dual-photon absorptiometry the beams of photons at two energies are provided either by one radioisotope with two gamma-emissions or by a combination of two radioisotopes. The detection systems are based on scintillation crystals, usually NaI. Theoretically, the optimum energies for the low and high energy beams are 35–45 keV and 200–800 keV respectively (Roos & Skoldborn, 1974). The lower energy provides high contrast between bone and soft tissue while mass attenuation at the higher energy is related to tissue density. The most common source used for DPA is ^{153}Gd, which emits gamma-rays with energies 44 keV and 100 keV. Combinations of radioisotopes that have been used include ^{125}I/^{241}Am (27.4 keV, 59.6 keV) and ^{241}Am/^{137}Cs (59.6 keV, 662 keV).

In recent years the use of radioisotopes has been superseded by X-ray technology. In DXA instruments the two collimated energy beams are produced in one of two ways, depending on the manufacturer (Sartorius & Resnick, 1988; Arai *et al.*, 1990). In the first method, X-rays with a spectrum of energies are produced by a constant-voltage X-ray generator and the beam is separated into high- and low-energy regions using k-edge filters of cerium or samarium. Current examples of this type of instrument include the Norland XR-26 (tube voltage 100 kV; peak beam energies 46.8 keV, 80 keV; dual NaI scintillation detection) and Lunar DPX (tube voltage 76 kV; peak beam energies 38 keV, 70 keV; single NaI scintillation detection). The second method involves a switching-pulse system that rapidly alternates the voltage of the X-ray generator producing two beams of high and low energy virtually simultaneously. An internal calibration wheel corrects for any small fluctuations caused by this method of beam generation and compensates for beam hardening effects. An example of this type of instrument is the Hologic QDR-1000 (tube voltages 70 kV, 140 kV; peak energies 43 keV, 110 keV; detection by $CdWO_4$ scintillation). DXA instrumentation is an improvement over DPA in that scan times are faster, imaging is better, precision is improved and the purchase, replacement and disposal of radioactive sources is not required.

Advantages and limitations of absorptiometry

Radiation exposure

One of the major advantages of absorptiometry compared with other techniques for assessing bone mineral content is that radiation exposure is low. The effective dose equivalent per measurement (EDE, a measure which takes into account surface exposure and penetrating power of the energy beam together with the depth and vulnerability of exposed tissues) depends on both the scan site and the instrument used, but values are generally within natural background radiation levels (Table 1.1; Kalendar, 1992). As a consequence, absorptiometry can be used to study healthy children and pre-menopausal women, as well as other, less vulnerable population groups, whereas methods such as neutron activation analysis and computed tomography are unsuitable. An exception is the measurement of the lumbar spine using DPA, as the EDE for pre-menopausal women is comparatively high because of the presence of the ovaries in the scan volume.

The surface exposure per measurement varies considerably between instruments and it is important for separate dosimetric evaluations to be made on each machine. Table 1.1 gives EDE estimates for pre-menopausal women measured on a number of instruments currently in use at the Dunn Nutrition Unit, Cambridge, together with some examples of EDE from exposure to other radiation sources. In general, the EDE values for men and for post-menopausal women are lower than those for pre-menopausal women (Kalendar, 1992). Although the radiation risks are low, it is prudent not to scan women during pregnancy.

Table 1.1. *Examples of effective dose equivalents (EDE) received during absorptiometry and from other sources in premenopausal women*

Radiation source	Instrument	EDE (μSv)
SPA midshaft radius	Norland 2780	0.04
SPA distal radius	Lunar SP2	0.18
DPA lumbar spine	Novolab 22A	20
DPA hip	Novolab 22A	2
DXA lumbar spine	Hologic QDR 1000/w	4
DXA hip	Hologic QDR 1000/w	4
DXA whole-body	Hologic QDR 1000/w	5
Daily background in Cambridge		6
Transatlantic return flight		80
Spinal radiograph		700
Cardangiographic examination		10 000

Data are from independent dosimetry of instruments used at the MRC Dunn Nutrition Unit (A. Prentice, unpublished data) and from Kalendar (1992). The effective dose equivalent (EDE) takes into account the surface radiation exposure, the penetrating power of the energy beam, and the depth and vulnerability of tissues irradiated.
SPA, single-photon absorptiometry; DPA, dual-photon absorptiometry; DXA, dual-energy X-ray absorptiometry.

Convenience

Absorptiometric techniques are very convenient as the patient/volunteer is required only to sit or lie still for the duration of the scan. Scan times on the latest instruments are much faster than with the older machines, with a whole-body scan typically taking 10–20 min and a lumbar spine scan taking 5–10 min. The nature of the technique means that absorptiometry can be used when other methods would be impossible. An example is the difficulty of using hydrodensitometry for body fat determinations when dealing with sick or disabled individuals. The necessity to keep still, however, can be a problem for some individuals, especially children and infirm people.

Precision and accuracy

The precision of absorptiometry is high and is superior to earlier methods for measuring bone mineral. The coefficients of variation of repeated bone scans across phantoms are about 1–2%, or less with the latest DXA instruments. The precision of repeated measurements is poorer when people are scanned, particularly if they are re-positioned each time or if they are osteoporotic, but generally within-day reproducibility is 3–5%. In addition, long-term reproducibility is good, especially with X-ray absorptiometry, which does not have the problems of source strength changes that inevitably effect long-term precision in SPA and DPA (Dunn *et al.*, 1987). Such high precision facilitates the longitudinal monitoring of individuals, although, even in osteoporosis, changes in bone mass can be sufficiently slow that measurements have to be spaced by several months or years for significant increments to be detectable (Davis *et al.*, 1991). The precision of fat and lean body mass determinations is also good and generally superior to other body composition methods such as hydrodensitometry and skinfold thickness measurements (e. g. Pritchard *et al.*, 1993). The precision of all absorptiometric measurements, however, is affected by poor positioning and excessive movement, and this can be a particular problem with children, disabled or anxious people.

The accuracy of absorptiometry is more problematical and, currently, there are a number of issues which give rise to concern and which remain unresolved. Accuracy can be affected by choice of calibrating materials, assumptions made in the computer programming about bone-edge detection and intra-osseous fat, the depths of tissues in the scan path, the non-uniformity of soft tissues overlying bone, and differences in clothing and bedding (Farrell & Webber 1989; Nord & Payne, 1990; Tothill & Pye 1990; Mazess *et al.*, 1991; Jonson, 1993). As a consequence, both absolute and relative values can differ substantially between manufacturers and between instruments. Use of different versions of computer software can affect results and updates of pre-existing programs necessitate re-analysis of data. This is a particular problem in longitudinal assessments or when comparisons are made between individuals measured on different machines. As a result, it is important that both the instrument and the software version used for analysis are specified in published accounts of absorptiometric studies. Possibly the greatest problems with accuracy are found with fat and lean tissue mass estimates, especially at the extremes of tissue depth (equivalent to >10 cm or <25 cm water), such as are encountered in children, anorexics, lean male athletes and the obese. At present there have been comparatively few validation experiments of the accuracy of either bone or body composition measurements by cadaver analysis and more are necessary before we can be confident about the accuracy of absorptiometry.

Bone measurements

One of the major limitations of bone measurements made by absorptiometry is that the result represents an integration of absorption over all elements within the bone envelope, such as the medullary cavity, and is not confined to osseous tissue *per se*. Thus a thin annulus of heavily mineralised bone in the forearm potentially could have the same measured bone mineral content as a wider annulus of osteoporotic bone with identical external dimensions. Similarly, absorptiometry cannot differentiate between trabecular and cortical bone, nor exclude abnormalities, such as osteophytes, crush fractures or calcifications, which interfere with the measurement of bone mineral content. Other methods, such as computed tomography and X-radiography are necessary if structural information about bone is required.

Interpretation of bone measurements by absorptiometry

Expression of bone data

The results from SPA are expressed as bone mineral content (BMC), which represents the mass of mineral in a slice of bone, 1 cm deep in the axial direction, and has the unit grams per centimetre. Results from DPA and DXA are expressed as bone mass per anatomical feature, such as spinal vertebrae or neck of femur, and the unit is grams. Confusingly, these values are also referred to as bone mineral content (BMC), although recently it has been suggested that the term bone mineral mass (BMM) should be adopted (Jonson, 1993).

To add to the confusion, BMC values from both single- and dual-energy absorptiometry are frequently divided by the width or area of the bone in the scan to give a figure referred to as the bone mineral density (BMD), with units grams per square centimetre. This value is not a true bone density, partly because it is only an areal rather than a volumetric measure, and partly because absorptiometry cannot distinguish between osseous and non-osseous tissues within the area designated as bone.

The interpretation of BMD is further complicated by the fact that BMC is not necessarily directly proportional to bone width (BW) or bone area (BA). In many population groups, a significant correlation remains between BMD values and BW (or BA), demonstrating that calculation of BMD does not completely adjust the variation in BMC due to differences in bone size. In a recent SPA study, for example, BMC at the radial shaft of young children was shown to be proportional, after age adjustment, to BW raised to the power 0. 77 (Prentice *et al.*, 1990). This shows that, in this case, the use of BMD to adjust for differences in bone size would artificially underestimate apparent bone density in those children with the widest bones at each age.

The purpose of calculating BMD is to facilitate the comparison of bone mineral measurements between individuals of different bone and body sizes. In practice, correlations often remain not only between BMD and BW (or BA) but also with indices of body size, such as body weight and height. A number of different approaches have been proposed for adjusting BMC or BMD data for body size. Many involve use of simple indices such as ideal weight-for-height (weight/height) or body mass index (BMI, weight/height^2). However, in most data sets, BMC and BMD are both positively and independently correlated with body weight and height, and the use of indices in which weight and height act in opposite directions is inappropriate. A better solution is to adjust BMC values for BW (or BA), body weight and height separately using multiple regression analysis (Cole & Prentice, 1992). A particularly informative procedure is to transform these variables to natural logarithms prior to analysis as this allows power and proportional relationships to be examined (Prentice *et al.*, 1990; Cole & Prentice, 1992).

This begs the question as to why it is felt necessary to express the data as either BMD or BMC adjusted for bone and body size. For technical reasons, BMD is a more precise measurement than BMC, as any movement during the scan will affect both BMC and bone width or area. For long-term monitoring of individual patients, use of BMD may be preferable as it removes some of the uncertainties due to measurement error. However, in osteoporosis screening the relevant variable is probably BMC as the strength of bone is strongly correlated to mineral mass (Dequeker, 1988; Geusens & Dequeker, 1988). Adjustments for bone and body size can be helpful, however, in assessing the relative BMC of individuals or groups of different size. Examples include comparing the bone mineral content of Blacks and Whites, or athletes and non-athletes, or assessing the importance of certain hormones in maintaining bone mineral. In these instances incomplete correction of BMC for size, for example by using either BMD alone or corrected for BMI, could lead to spurious relationships emerging, as the residual influence of size on BMC could result in apparent correlations with other size-related variables (Cole & Prentice, 1992). The use of multiple regression analysis with BMC as the dependent variable, as described above, avoids the complexities which are introduced by using derived indices such as BMD and BMI to adjust for size differences.

Reference data

In general, osteoporosis screening and diagnosis are undertaken by comparing an individual's BMC or BMD value with data from a reference population. The patient's results are expressed relative to the sex-specific reference population either at the subject's age or for young adults. Clinical interpretation is subjective but typically individuals are considered at increased fracture risk if their values are less than 80% or 2 standard deviations (SD) below the reference mean.

At present there are difficulties with the use of reference data. Firstly, the reference distribution is provided by the manufacturers and may not be pertinent to local population groups. In addition, differences between reference databases and instrumentation can lead to anomalies where a patient appears normal on one machine and osteoporotic on another (Laskey *et al.*, 1992), thereby affecting the diagnosis and the management of the patient.

Reversible calcium space

Bone remodelling is a dynamic process whereby resorption at specific sites is followed, after a period of time, by bone formation. In healthy young adults these two processes are linked so that, ultimately, total bone mass is unaffected. When the rate of bone turnover decreases, bone formation continues at pre-existing resorption sites, while fewer new resorption sites are formed. Eventually bone formation decreases to match resorption, but the process can take several months (Melsen & Mosekilde, 1988). Absorptiometry is a static measurement which provides a 'snapshot' of BMC at any one moment. If bone turnover decreases with no overall change in bone volume, BMC values will rise by a few per cent because of the smaller number of resorption cavities. BMC values continue to rise for some time until a new steady state is achieved because of the time lag between resorption and formation (Parfitt, 1980; Kanis, 1991). Similarly, a decrease in reversible calcium space caused by an increase in bone turnover will lead to a decrease in measured BMC.

Such artefacts caused by changes in reversible calcium space potentially could confound interpretation of absorptiometric studies. In a patient with osteoporosis, an observed rise in BMC after drug treatment might be due to reduced bone turnover rather than to increased bone mass. In this instance, the decrease in turnover might of itself be beneficial, as bone loss would be diminished, and the exact interpretation of the BMC change could be regarded as academic. In other cases, however, a proper understanding of the mechanism could alter the conclusions reached. An example is provided by a recent study of calcium supplementation during childhood which involved pre-pubertal twins (Johnston *et al.*, 1992). Those twins who received a calcium supplement of about 700 mg Ca/day on top of a dietary intake of about 900 mg Ca/day were shown to have significantly higher BMC (BMD) values than their co-twins after 3 years. This has been interpreted as showing that the calcium-supplemented twins had a greater rate of bone acquisition and that possibly the calcium requirements of pre-pubertal children are higher than previously thought. However, the bone turnover of the supplemented twin, as shown by circulating osteocalcin levels, was significantly lower than that of the co-twin, suggesting that the observed differences in BMC were, at least in part, due to effects on reversible calcium space, and may not have been due to differences in bone mass. Consequently, the likely impact of the calcium supplementation on the development of peak bone mass is uncertain and must await long-term follow-up. Indeed, it is not too fanciful to hypothesise that the twin with the lower bone turnover might have slower bone development and that this might result in a lower peak bone mass. It is obviously essential that more research is undertaken to understand the relevance and importance of reversible calcium space in the interpretation of absorptiometric data, and in future it may be worth while including measures of bone turnover in research protocols involving absorptiometry.

Summary

It is now possible to measure the bone mineral content and body composition of a wide range of individuals, in health and during illness, using a variety of instruments based on absorptiometry. Single- and dual-energy absorptiometric measurements are quick and easy to make, are accompanied by only a very small radiation exposure and are highly precise. However, there are still considerable concerns about the accuracy of these techniques, especially at extremes of tissue depth, and interpretation of the data is complex. Despite substantial advances in instrumentation and understanding in recent years, there are still many pitfalls for the unwary and absorptiometers cannot be treated as 'black-box' machines that will produce meaningful results in unskilled hands and in all circumstances. The applications of absorptiometry, especially DXA, are likely to increase in the future as the instruments become more widely available in research laboratories.

References

Arai, H., Ito, K., Nagao, K. & Furutachi, M. (1990). The evaluation of three different bone densitometry systems: XR-26, QDR-1000 and DPX. *Image Technology and Information Display*, **22**, 1–6.

Cameron, J.R. & Sorenson, J. (1963). Measurement of bone mineral *in vivo*: an improved method. *Science*, **142**, 230–2.

Cameron, J.R., Grant, R. & MacGregor, R. (1962). An improved technique for the measurement of bone mineral content *in vivo*. *Radiology*, **78**, 117.

Cole, T.J. & Prentice, A. (1992). Bone mineral measurements. *British Medical Journal*, **305**, 1223–4.

Davis, J.W., Ross, P.D., Wasnich, R.D., MacLean, C.J. & Vogel, J.M. (1991). Long-term precision of bone loss rate measurements among postmenopausal women. *Calcified Tissues International*, **48**, 311–18.

Dequeker, J. (1988). Calcified tissues: structure-function relationships. In *Calcium in Human Biology*, ed. B.E.C. Nordin, pp. 209–40. London: Springer-Verlag.

Dunn, W.L., Kan, S.H. & Wahner, H.W. (1987). Errors in longitudinal measurements of bone mineral: effect of source strength in single and dual photon absorptiometry. *Journal of Nuclear Medicine*, **28**, 1751–7.

Farrell, T.J. & Webber, C.E. (1989). The error due to fat inhomogeneity in lumbar spine bone mineral measurements. *Clinical Physics and Physiological Measurement*, **10**, 57–64.

Geusens, P. & Dequeker, J. (1988). Should bone mineral content measurements be corrected for skeletal size? In *Bone Mineral Measurement by Photon Absorptiometry*, ed. J. Dequeker, P. Geusens & H.W. Wahner, pp. 96–104. Leuven: Leuven University Press.

Johnston, C.C., Slemenda, C.W. & Melton, L.J. (1991). Clinical use of bone densitometry. *New England Journal of Medicine*, **324**, 1105–9.

Johnston, C.C., Miller, J.Z., Slemenda, C.W., Reister, T.K., Hui, S., Christian, J.C. & Peacock, M. (1992). Calcium supplementation and increases in bone mineral density in children. *New England Journal of Medicine*, **327**, 82–7.

Jonson, R. (1993). Mass attenuation coefficients, quantities and units for use in bone mineral determinations. *Osteoporosis International*, **3**, 103–5.

Kalendar, W. A. (1992). Effective dose values in bone mineral measurements by photon absorptiometry and computed tomography. *Osteoporosis International*, **2**, 82–7.

Kanis, J.A. (1991). Calcium requirements for optimal skeletal health in women and osteoporosis. *Calcified Tissues International* (Supplement) **49**, S33–41.

Laskey, M.A., Crisp, A.J., Cole, T.J. & Compston, J.E. (1992). Comparison of the effect of different reference data on Lunar DPX and Hologic QDR-1000 dual-energy X-ray absorptiometers. *British Journal of Radiology*, **65**, 1124–9.

Mazess, R.B. (1971). Estimation of bone and skeletal weight by direct photon absorptiometry. *Investigative Radiology*, **6**, 52–60.

Mazess, R.B., Trempe, J.A., Bisek, J.P., Hanson, J.A. & Hans, D. (1991). Calibration of dual-energy X-ray absorptiometry for bone density. *Journal of Bone and Mineral Research*, **6**, 799–806.

Melsen, F. & Mosekilde, L. (1988). Calcified tissues: cellular dynamics. In *Calcium in Human Biology*, ed. B.E.C. Nordin, pp. 187–208. London: Springer-Verlag.

Nord, R.H. & Payne, R.K. (1990). Standards for body composition calibration in DEXA. In *Current Research in Osteoporosis and Bone Mineral Measurement*, ed. E.F.J. Ring, pp. 27–8. London: The British Institute of Radiology.

Parfitt, A.M. (1980). Morphologic basis of bone mineral measurements: transient and steady state effects of treatment in osteoporosis. *Mineral and Electrolyte Metabolism*, **4**, 273–87.

Peppler, W.W. & Mazess, R.B. (1981). Total body bone mineral and lean body mass by dual-photon absorptiometry. I. Theory and measurement procedure. *Calcified Tissues International*, **33**, 353–9.

Prentice, A., Laskey, M.A., Shaw, J., Cole, T.J. & Fraser, D.R. (1990). Bone mineral content of Gambian and British children aged 0–36 months. *Bone and Mineral*, **10**, 211–24.

Pritchard, J.E., Nowson, C.A., Strauss, B.J., Carlson, J.S., Kaymakci, B. & Wark, J.D. (1993). Evaluation of dual energy X-ray absorptiometry as a method of measurement of body fat. *European Journal of Clinical Nutrition*, **47**, 216–28.

Roos, B.O. & Skoldborn, H. (1974). Dual photon absorptiometry in lumbar vertebrae. I. Theory and method. *Acta Radiologica*: Therapeutics, **13**, 266–80.

Sartoris, D.J. & Resnick, D. (1988). Digital radiography may spark renewal of bone densitometry. *Diagnostic Imaging*, January, 145–51.

Sorenson, J.A. & Cameron, J.R. (1967). A reliable *in vivo* measurement of bone-mineral content. *Journal of Bone and Joint Surgery* [*Am*], **49A**, 481–97.

Tothill, P. (1989). Methods of bone mineral measurement. *Physics in Medicine and Biology*, **34**, 543–72.

Tothill, P. & Pye, D.W. (1990). Errors due to fat in lateral spine DPA. In *Current Research in Osteoporosis and Bone Mineral Measurement*, ed. E.F.J. Ring, p. 19. London: The British Institute of Radiology.

Wahner, H.W. & Riggs, B.L. (1986). Methods and application of bone densitometry in clinical diagnosis. *CRC Critical Reviews in Clinical Laboratory Sciences*, **24**, 217–33.

Wahner, H.W., Dunn, W.L. & Riggs, B.L. (1984). Assessment of bone mineral. Part 2. *Journal of Nuclear Medicine*, **25**, 1241–53.

2 *In vivo neutron activation analysis: past, present and future*

S.J.S. RYDE

Introduction

It is 30 years since the first measurements of the elemental composition of the human body were undertaken by Anderson *et al.* (1964) using the technique of *in vivo* neutron activation analysis (IVNAA). Their work followed a report (Hoffman & Hempelmann, 1957) of two nuclear reactor accidents in 1945–6 during which ten persons were exposed to bursts of radiation (fast neutrons and gamma-rays). A measure of the serum ^{24}Na activity induced in the body was used to estimate the neutron intensity to which the subjects had been exposed. It was subsequently realised that a controlled irradiation with neutrons of known intensity could be used as an investigative tool.

The IVNAA technique involves the irradiation of the total or partial body by a beam of neutrons and the detection of characteristic gamma-rays arising from neutron interaction with the nuclei of the element or elements of interest. Although the principle of the technique is simple, the factors that determine the successful measurement of a particular element *in vivo* can be complex. Nevertheless, IVNAA has been successfully developed at several centres over the last three decades and gained increasing acceptance as a clinically useful, although specialised measurement technique. During this time considerable research has been undertaken to optimise the measurement of both bulk (e.g. Ca, C, Cl, H, N, Na, O and P) and trace (e.g. Al, Cd, Cu, Fe and Si) elements. However, the measurements of calcium, cadmium and nitrogen have been the most widely applied.

A number of review articles on the techniques and clinical applications of IVNAA have been published (Cohn, 1980, 1981; Chettle & Fremlin, 1984; Beddoe & Hill, 1985; Cohn & Parr, 1985) and these provide excellent sources of information, including extensive reference lists. More recently, proceedings have been published from the '*In Vivo* Body Composition Studies' series of symposia (Ellis *et al.*, 1987; Yasumura *et al.*, 1990; Ellis & Eastman, 1993) and these too provide a valuable source of information on developments in IVNAA (and other techniques of body composition).

This chapter, not intended for experienced practitioners of IVNAA, will describe briefly the principles upon which the technique is based and the basic measurement methods utilised in recent years, and will consider the future of the technique. The emphasis will be on the measurement of total-body nitrogen because this element appears to offer the greatest clinical potential for the future.

Basic principles of IVNAA

The technique of IVNAA is based upon the detection and measurement of the various radiations emitted as a result of fast- or thermal-neutron interactions with elements in the body. The products of these interactions may be radionuclides having specific half-lives and gamma-ray emissions which are measured after the end of the irradiation (termed delayed-gamma neutron activation analysis, DGNAA), or they may be excited states of stable nuclides the characteristic 'prompt' gamma-rays of which must be measured during the neutron irradiation because the excited compound nuclei rapidly (within, say 10^{-12} s) return to their ground state (termed prompt-gamma neutron activation analysis, PGNAA). Radiation detectors placed around the body measure the energy and intensity of the characteristic emissions which, after calibration by comparison with similar emissions from a suitable human-like phantom, enables the amount of the element of interest to be calculated. The nuclear reaction and mode of de-excitation that predominates is dependent not only on the nuclide involved but also on the neutron energy. For a more complete account of the physics relevant to neutron activation analysis the reader is referred to the work of De Soete *et al.* (1972).

Fast-neutron activation

In the case of activation with fast neutrons, de-excitation of the highly excited compound nucleus predominantly occurs by scattering, the emission of nucleons (protons, alpha particles and neutrons) and the emission of gamma radiation.

Inelastic scattering, and the production of characteristic gamma radiation which is usually prompt, is not widely used in *in vivo* analysis. It does, however, have one important application in the measurement of carbon (Kyere *et al.*, 1982; Kehayias *et al.*, 1987; Sutcliffe *et al.*, 1990). Reactions involving emission of nucleons are also little used; the most notable exception has been the measurement of nitrogen by the $^{14}N(n,2n)^{13}N$ reaction (although, as will been seen later, this is no longer the reaction of choice for the determination of nitrogen *in vivo*).

Thermal-neutron activation

There are many elements in which de-excitation can occur only by gamma emission. This type of reaction, radiative capture, is the most widely used of all for *in vivo* analysis. Radiative capture predominates at low neutron energies because the reaction cross-section is approximately proportional to the inverse of the neutron speed.

Neutron interactions in tissue

Fast neutrons incident on the body are slowed predominantly by inelastic and elastic scattering reactions with the body tissues until thermal equilibrium is reached. During elastic scattering the neutron transfers to the nucleus a proportion of this kinetic energy. Hydrogen, which accounts for about 62% of the nuclei in the body, is by far the most important constituent of the body for the slowing down or moderation of fast neutrons. Fast neutrons of a few MeV energy are moderated to thermal energies in a few microseconds. Once thermalised, radiative capture reactions may proceed for about 200 μs as the thermal neutrons diffuse throughout the tissues, after which time any prompt gamma-rays produced will decrease substantially.

In vivo reactions

The choice of reaction for a given elemental measurement depends upon many parameters. Foremost are the nuclear parameters such as isotopic abundance, neutron reaction cross-section (probability of interaction) and the characteristics of the emitted radiation, since these cannot be changed. However, other parameters, for example the neutron energy spectrum and fluence, can be changed. Increasing neutron energy introduces greater potential for utilising reactions having fast-neutron thresholds, i.e. the fast neutron-induced reaction used in the measurement of carbon as noted previously.

Other factors, including the availability of a suitable gamma-ray spectroscopy system, also influence the choice of neutron reaction utilised and what elements are measurable. Prompt-gamma techniques, for example, place considerable demands upon the spectroscopy system because of the high counting rates involved. Furthermore, because a typical prompt gamma-ray spectrum contains many peaks it is sometimes necessary to use semi-conductor-type detectors. The exceptions to this occur in the measurement of nitrogen, hydrogen, chlorine and carbon, for which large-volume NaI(Tl) scintillation detectors are used. However, the alternative delayed-gamma technique requires the availability of a whole-body counter if total-body measurements are to be undertaken.

A comparison of the prompt and delayed techniques indicates that there may be disadvantages associated with the latter, namely (1) the constraints imposed by the fact that it is necessary to adhere to periods of irradiation, transfer and counting because of the half-life of the particular isotope, (2) the fact that there are fewer elements measurable since the product isotope must be radioactive, and (3) the loss of useful data (gamma-ray counts) throughout the period of irradiation and transfer. The major advantages of the delayed-gamma technique are that (1) there are no spectroscopy problems associated with high counting rates, (2) there is no neutron-induced damage to the gamma-ray detector, and (3) the gamma-ray spectrum is less complex, although gamma-rays of similar energy may still interfere with one another.

Table 2.1 lists the range of elements and reactions that have been used in clinical studies in recent years. Many studies have been undertaken to assess the feasibility of measuring a range of elements, using many different reactions; however, only those reaction that have found clinical utility are listed. Other reaction parameters such as whether it is fast- or thermal-neutron induced, produces prompt or delayed gamma-rays, and the radioactive half-life (if applicable) are given.

Neutron sources

A number of neutron sources have been used for IVNAA including cyclotrons (Biggin *et al.*, 1972; Kennedy *et al.*, 1982), 14 MeV D,T neutron generators (Haywood *et al.*, 1981; Sharafi *et al.*, 1983; Kehayias *et al.*, 1990), reactor neutrons (Ellis & Kelleher, 1987), and radionuclide sources such as ^{238}Pu,Be (Cohn *et al.*, 1972; McNeill *et al.*, 1973; Vartsky *et al.*, 1979a; Beddoe *et al.*, 1984*b*), and ^{252}Cf (Allen *et al.*, 1987; Larsson *et al.*, 1987; Ryde *et al.*, 1987; Mackie *et al.*, 1988; Krishnan *et al.*, 1990; Baur *et al.*, 1991). There are several factors that govern the choice of source, including (1) the element to be activated, (2) the spatial uniformity of measurement required, (3) radiation dose, (4) source complexity and size, and (5) cost.

The usual choice of neutron source if reactions with a threshold energy (i.e. carbon) are to be utilised is the 14 MeV D,T neutron generator. This source has advantages that include the ability to provide a pulsed beam and excellent tissue penetration. However, there are also disadvantages: for example, the neutron beam output must be monitored, the generator requires ancillary electronics to operate, and the radiation dose per incident

Table 2.1. *Body elements measured by* in vivo *neutron activation analysis in recent years*

Element	Weight[a] in body (g)	Reaction	Fast or thermal neutron induced	Half-life (s)	Gamma-ray energy (Mev)	Reference[b]
C	16000	$^{12}C(n,n'\gamma)^{12}C$	F	Prompt	4.439	1, 2, 3
H	7000	$^{1}H(n,\gamma)^{2}H$	T	Prompt	2.223	4, 5, 6
N	1800	$^{14}N(n,\gamma)^{15}N$	T	Prompt	10.829	4, 5, 6, 7, 8, 9, 10
		$^{14}N(n,2n)^{13}N$	F	598	0.511	11, 12
Ca	1000	$^{48}Ca(n,\gamma)^{49}Ca$	T	523	3.084	13, 14, 15, 16, 17
		$^{40}Ca(n,\gamma)^{41}Ca$	T	Prompt	6.420	18
P	780	$^{31}P(n,\alpha)^{28}Al$	F	134	1.779	13, 15, 17
Na	100	$^{23}Na(n,\gamma)^{24}Na$	T	54072	2.754	13, 15, 19
					1.369	20
Cl	95	$^{35}Cl(n,\gamma)^{36}Cl$	T	Prompt	6.111	21, 22
					8.579	23
		$^{37}Cl(n,\gamma)^{38}Cl$	T	2238	2.168	13, 15, 19
					1.642	
Al	0.061	$^{27}Al(n,\gamma)^{28}Al$	T	134	1.779	24, 25
Cd	0.024	$^{113}Cd(n,\gamma)^{114}Cd$	T	Prompt	0.559	26, 27

[a] ICRP (1975).
[b] This is not an exhaustive reference list and in general only the more recent references are given since these cite much of the earlier work.
1, Kyere *et al.* (1982); 2, Kehayias *et al.* (1990); 3, Sutcliffe *et al.* (1990); 4, Vartsky *et al.* (1984); 5, Ryde *et al.* (1989); 6, Beddoe *et al.* (1984*a*); 7, Mackie *et al.* (1990); 8, Krishnan *et al.* (1990); 9, Baur *et al.* (1991); 10, Larsson *et al.* (1987); 11, Haywood *et al.* (1981); 12, McCarthy *et al.* (1980); 13, Dilmanian *et al.* (1990); 14, Kennedy *et al.* (1982); 15, Sharafi *et al.* (1983); 16, Harrison *et al.* (1990); 17, Glaros *et al.* (1987); 18, Ryde *et al.* (1994); 19, Kennedy *et al.* (1983); 20, Spinks *et al.* (1980); 21, Beddoe *et al.* (1987); 22, Ryde *et al.* (1990*a*); 23, Mitra *et al.* (1993); 24, Ellis & Kellcher (1987); 25, Wyatt *et al.* (1993); 26, Morgan *et al.* (1990); 27, Franklin *et al.* (1990).

neutron is higher than obtained using lower energy sources such as ^{238}Pu,Be (Cohn *et al.*, 1973).

Radionuclide sources are the most popular choice for the majority of IVNAA measurements because of their simplicity, predictable output, portability and compact size. However, a mean neutron energy ($\sim$4.4 MeV for ^{238}Pu,Be and $\sim$2.1 MeV for ^{252}Cf) that is lower than the D,T generator leads to comparatively poorer thermal-neutron fluence spatial uniformity in tissue (Cohn *et al.*, 1973). Spatial uniformity is important for *in vivo* analysis to ensure that, as far as is practicable, activation of the element of interest is independent of the position in the body. In practice a combination of fast-neutron beam pre-moderation, bilateral irradiation and careful positioning of the gamma-ray detectors around the body (necessary

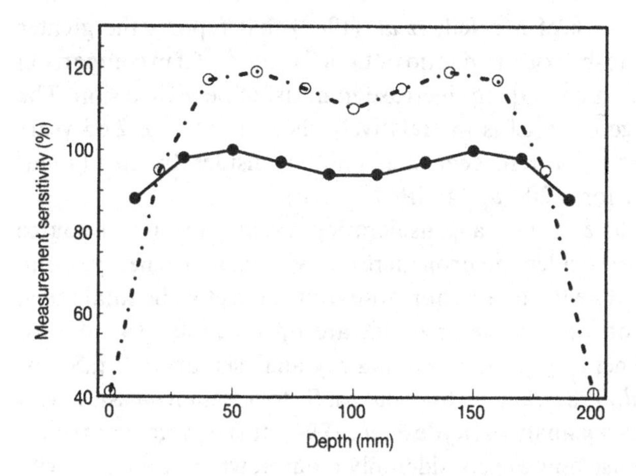

Fig. 2.1. Combined thermal-neutron activation and detection uniformity (filled circles) of 8.99 MeV gamma-rays with depth in a 200 mm deep water phantom irradiated bilaterally by ^{252}Cf neutrons. The profile from activation alone (open circles) is shown for comparison.

because the emitted gamma-rays will have different paths to the detectors) can achieve acceptable measurement uniformity of ± 6.4, ± 4.0 and $\pm 9.0\%$ for water thicknesses of 150, 200 and 250 mm, respectively, using ^{252}Cf and 8.99 MeV gamma-rays (Ryde *et al.*, 1989). A marginally better uniformity is obtained using ^{238}Pu,Be (Vartsky *et al.*, 1979*a*). Fig. 2.1 illustrates a typical measurement profile obtained using a ^{252}Cf neutron source to irradiate a 200 mm thick water phantom.

Neutron beam penetration and measurement uniformity may also be influenced by other factors such as neutron collimator length, aperture size and fabrication material (Morgan *et al.*, 1981; Mountford 1982; Ryde *et al.*, 1987). In general, therefore, the design of IVNAA apparatus requires great care to ensure that the combined effect of spatial variation in the activating thermal-neutron fluence and emitted gamma-rays is minimised and an optimum measurement is achieved.

A major benefit of utilising a source with a low mean neutron energy is that, when compared with a source of higher energy, it produces a greater thermal-neutron fluence per unit of incident radiation dose thereby increasing the yield of radiative capture gamma-rays. Comparison of ^{252}Cf with ^{238}Pu,Be has shown that the neutrons from the former source generate nearly 40% more thermal fluence per incident dose (Morgan *et al.*, 1981). This observation has established ^{252}Cf as the neutron source of choice for the measurement of elements that can proceed via thermal-neutron reactions, such as nitrogen and chlorine. This conclusion has been further supported by the work of McNeill *et al.* (1989) that reports the greater gamma-ray signal-to-background ratio obtained using ^{252}Cf in comparison with ^{238}Pu,Be with a consequent increase in measurement precision. The major disadvantage of ^{252}Cf is the relatively short half-life of 2.65 years which necessitates regular replacement at a not inconsiderable cost (about £17 000 in the UK for a 200 μg (4 GBq) source).

The radiation dose is also a consideration when choosing a source because the dose per incident neutron increases with neutron energy (Cohn *et al.*, 1973). Typical neutron radiation dose equivalents for the total body, assuming a neutron quality factor of 10, are up to about 0.4 mSv for nitrogen measurement by prompt gamma-ray analysis, up to 2 mSv for carbon (Kyere *et al.*, 1982), and up to about 5 mSv for calcium measurement by prompt gamma-ray analysis (Ryde *et al.*, 1994). It is of note that typical radiation doses in use now are considerably reduced when compared with those of 10 mSv or more commonly used 10–20 years ago (Chamberlain *et al.*, 1968, Nelp *et al.*; 1970, Kennedy *et al.*, 1982; Sharafi *et al.*, 1983). The radiation dose equivalent given above for nitrogen falls within the typical variations of the natural background radiation. Above all else, the radiation dose to the patient should be considered in the light of the benefit likely to be derived from the measurement.

Instrumentation and procedures for IVNAA

Irradiation apparatus

A variety of instruments have been used for clinical measurements. The early instruments were often adapted from existing resources (intended for nuclear physics research or radiotherapy research) such as cyclotrons (Biggin *et al.*, 1972, Spinks *et al.*, 1977, Kennedy *et al.*, 1982) and reactor facilities (Boddy & Alexander, 1967), or were purpose built close to nuclear physics research laboratories because of the readily available expertise (Cohn & Dombrowski, 1971; Boddy *et al.*, 1974). In recent years, however, there has been a desire to ensure that instruments are both affordable and clinically accessible. The latter is satisfied in practice by using a radioisotopic neutron source such as ^{252}Cf or ^{238}Pu,Be, or a compact 14 MeV D,T generator, and by having the instrument appropriately sited within the hospital environment to facilitate clinical use.

The majority of instruments built within the last 10–15 years have been primarily for the measurement of total-body nitrogen or carbon by prompt-gamma-ray analysis. An exception has been the Swansea instrument that has a multi-element capability permitting the measurement of calcium, nitrogen and cadmium by PGNAA (Ryde *et al.*, 1987). Recently built

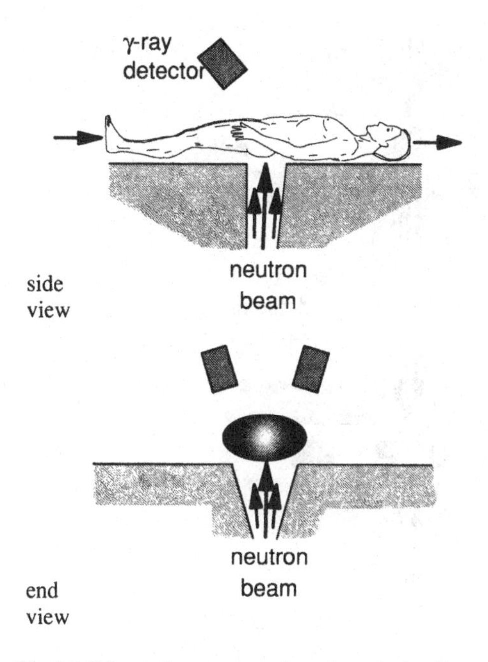

Fig. 2.2. Schematic representation of a typical unilateral irradiation geometry. The gamma-ray detectors may be positioned above or to the side of the subject.

instruments have generally followed one of two irradiation geometries: unilateral or bilateral irradiation.

The unilateral geometry (Fig. 2.2) provides irradiation of the subject from below the couch. The prompt gamma-rays are measured by detectors placed either above or to the side of the subject (or a combination of the two). Irradiation and detection on opposite sides of the subject (i.e. detectors placed above the subject) provides the most favourable conditions for uniformity of activation and detection when the subject is irradiated on one side alone. The subject can be scanned in both the prone and supine positions to improve the uniformity further if required. The bilateral geometry (Fig. 2.3) provides simultaneous bilateral irradiation by placing a neutron source both above and below the subject. For this configuration the detectors are placed to the side of the subject.

Studies with instruments utilising the unilateral irradiation geometry for the measurement of nitrogen include Vartsky *et al.* (1979*a*), Allen *et al.* (1987), Larsson *et al.* (1987), Ryde *et al.* (1987) and Stroud *et al.* (1990), and for the measurement of carbon include Kyere *et al.* (1982) and Kehayias *et al.* (1987) (noting that Kyere *et al.* irradiated from the side of the subject, not in the anterior–posterior direction). The bilateral geometry has been

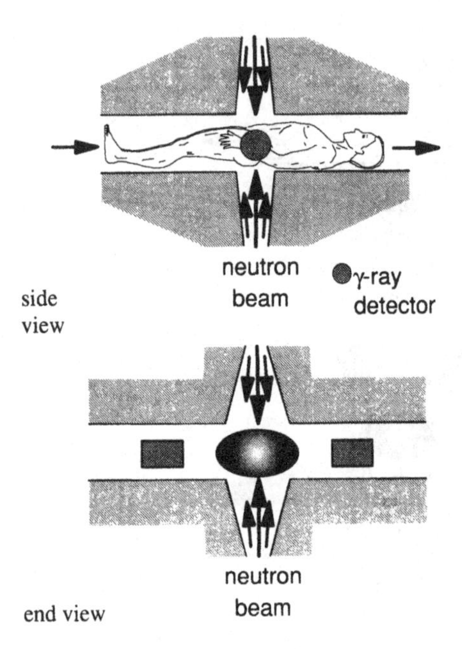

Fig. 2.3. Schematic representation of a typical simultaneous bilateral irradiation geometry.

utilised for the measurement of nitrogen by Mernagh *et al.* (1977), Beddoe *et al.* (1984*b*), Mackie *et al.* (1988) and Krishnan *et al.*, (1990). Some of the above instruments, primarily designed for the measurement of nitrogen, have also been used for the determination of chlorine (Beddoe *et al.*, 1987; Ryde *et al.*, 1990*a*; Blagojevic *et al.*, 1990; Mitra *et al.*, 1993) and carbon (Sutcliffe *et al.*, 1990).

There are advantages and disadvantages associated with both measurement geometries. Unilateral irradiation geometry offers a cost advantage because only one neutron source is used and less radiation shielding is needed. Furthermore, a wide choice of detector position is available (i.e. above or to the side of the subject) which can help to achieve an optimum measurement uniformity. The prime disadvantage of this geometry is the need to scan the subject both supine and prone if a bilateral irradiation is required. Such positioning might be difficult for seriously ill subjects who may have feeding tubes and monitoring equipment attached. Clearly, this is not a concern with the simultaneous bilateral irradiation geometry and is its major advantage. Other advantages and disadvantages of the simultaneous bilateral geometry tend to be the opposite of those given above, i.e. greater cost because of two sources and so forth. One further disadvantage of the bilateral geometry might arise, somewhat infrequently, from a claustrophobic effect with some subjects.

Gamma-ray detectors and spectroscopy

There are two types of gamma radiation detector used for IVNAA: the semiconductor and the scintillation crystal. The former detector offers very good energy resolution (typically 2 keV or better at 1332 keV) but poor efficiency, whereas the latter provides much better efficiency by a factor of 10 or more but has poorer energy resolution by a factor of about 40.

There are a variety of semiconductor type detectors available but the most useful for IVNAA is probably the n-type hyperpure germanium semiconductor configured in a coaxial geometry (Knoll, 1988). This detector must be cooled to liquid nitrogen temperature during use, although this presents only a minor nuisance in practice. The semiconductor detector is normally used only when the gamma-ray energy to be measured is close to other gamma-ray emissions in the spectrum that would otherwise interfere with the analysis. Examples of its application include the measurement of cadmium (Scott & Chettle, 1986) and the measurement of calcium (Ryde *et al.*, 1987).

The most common detector in use for IVNAA is the NaI(Tl) scintillation detector. It is universally used in whole-body radiation counters for the measurement of delayed gamma-rays and for the measurement of nitrogen and carbon by prompt gamma-ray analysis. Whole-body counters vary in complexity from simple shadow-shield designs containing two detectors to shielded rooms containing up to 32 crystals (Dilmanian *et al.*, 1990), although whole-body counters containing this large number of detectors are unusual and eight or fewer detectors are typical (Harrison *et al.*, 1990).

When scintillation detectors are used for PGNAA it is common to have only two or four detectors (nominal dimensions 152 mm diameter × 152 mm thick). Because these measurements are made during the neutron irradiation, the detectors must be shielded to reduce the number of fast and thermal neutrons reaching the scintillation crystal and so reduce the induced background counts. Nevertheless, gamma-ray counting rates can be high (30 000 counts per second or more) which requires the use of high-performance spectroscopy systems. Typical systems comprise a photomultiplier assembly, designed for high counting rates with a high-stability dividing circuit to maintain interdynode voltage and negative high voltage on the photocathode, a spectroscopy amplifier with facilities to adjust pulse shaping time and baseline restoration circuits, analogue-to-digital converter with short conversion time (say a mean conversion time of 3 μs or less) and a multi-channel analyser (now normally integrated with a personal computer). An independent spectroscopy system will often be required for each detector to ensure maximum throughput of gamma-ray counts.

Calibration

A measure of the absolute mass of an element is an important goal for IVNAA. Appropriate calibration is needed, therefore, to relate the gamma-ray counts arising from a particular element irradiated under known conditions to the amount of that element present in the body. This relationship is complex and dependent upon several factors including the neutron beam intensity, the nuclear characteristics of the element to be measured, the activation and detection uniformity and so forth. Thus each instrument requires a unique calibration that will involve extensive evaluation of phantoms containing a near tissue-equivalent solution of known composition and representing a range of body size (Cohn *et al.*, 1972; Williams *et al.*, 1978; Kennedy *et al.*, 1982; Kyere *et al.*, 1982; Beddoe *et al.*, 1984*b*; Ryde *et al.*, 1989; Sutcliffe *et al.*, 1990; Mackie *et al.*, 1990; Baur *et al.*, 1991; Mitra *et al.*, 1993). The most robust calibration is likely to involve inhomogeneous phantoms, especially for the measurement of elements such as calcium that are not uniformly distributed in the body.

A significant advance in the absolute PGNAA measurement of nitrogen *in vivo* was made by Vartsky and colleagues in Birmingham (Vartsky *et al.*, 1979*b*) when it was suggested that an 'internal standardisation' technique could be employed. This technique, now in widespread use, relies on measuring the ratio of nitrogen-to-hydrogen gamma-ray counts, since this is less dependent on body habitus than the nitrogen gamma-ray counts alone. It requires that the body mass of hydrogen, the internal standard, is assessed independently. This has been achieved by several methods, viz. (1) by assuming it is proportional to body mass (McNeill *et al.*, 1982), (2) from body fat (via skinfold anthropometry) and body mass (Vartsky *et al.*, 1979*b*), (3) from total-body water, body mass and fat (Vartsky *et al.*, 1979*a*), (4) from total-body water, body mass, fat, minerals and glycogen (Beddoe *et al.*, 1984*b*) and (5) from total-body water, body mass, fat and total body calcium giving the bone mineral ash (Vartsky *et al.*, 1984). Vartsky and colleagues (1979*b*) concluded that errors due to different irradiation and detection conditions, and to different shapes and positions of the subject, were reduced by a factor of 3 or more when using the hydrogen internal standardisation technique. The justification for the technique has been validated by Vartsky *et al.* (1979*b*) and subsequently by others (Knight *et al.*, 1986).

Internal standardisation by hydrogen has now been applied to the absolute measurement by PGNAA of other elements in addition to nitrogen – for example, chlorine (Beddoe *et al.*, 1987; Blagojevic *et al.*, 1990) and carbon (Sutcliffe *et al.*, 1990). More recently, the concept of internal standardisation has been extended to the measurement of calcium utilising the prompt gamma-ray technique; in this application the internal standard is total-body chlorine (Ryde *et al.*, 1990*b*; Evans *et al.*, 1994).

If sequential studies over time are to be performed on the same subject and changes rather than absolute amounts of the elemental content are to be calculated then the results of the analysis can be presented as gamma-ray counts and calibration is simplified. It is emphasised that this is true only whilst the body size and shape of the subject remain constant and the performance of the irradiation and detection apparatus does not change. In practice, even when undertaking sequential studies, phantoms are regularly measured to check the constancy of the latter and to provide normalisation if necessary.

Practical measurement procedures

Given a calibrated instrument, the measurement of body composition by IVNAA is not complicated. The measurement of nitrogen by PGNAA is typically performed as follows: (1) the subject is scanned between the shoulder and knee for a total time of about 30 min, (2) a prompt gamma-ray spectrum, such as that shown in Fig. 2.4, is obtained and the number of gamma-ray counts in the nitrogen and hydrogen gamma-ray regions of interest calculated, (3) a number of lateral and anterior–posterior dimensions of the subject are recorded (with the subject in the scanning position) in order to make corrections for the effect of body size on the nitrogen and hydrogen gamma-rays. Total-body nitrogen is then obtained from the corrected nitrogen and hydrogen counts and a model of body composition. Concurrent with the IVNAA procedure the subject is given a small (4 MBq) amount of tritiated water to drink. The tritiated water provides a measure of the tritiated-water space, and thus the total-body water, and is used in the model of body composition. Normally a four- or five-compartment model of body composition (for example protein (as 6.25 × nitrogen), water, fat, minerals and glycogen) is used (Vartsky *et al.*, 1984; Beddoe *et al.*, 1984*a*). It is to be noted that when using a model such as the five-compartment model above, the calculation of protein (nitrogen) is relatively insensitive to errors in the measurement of the other compartments. The measurement procedure for the analysis of carbon and chlorine by PGNAA is essentially identical to that for nitrogen.

The measurement procedure using DGNAA is somewhat different to that outlined above. During DGNAA the neutron dose-rate is not limited

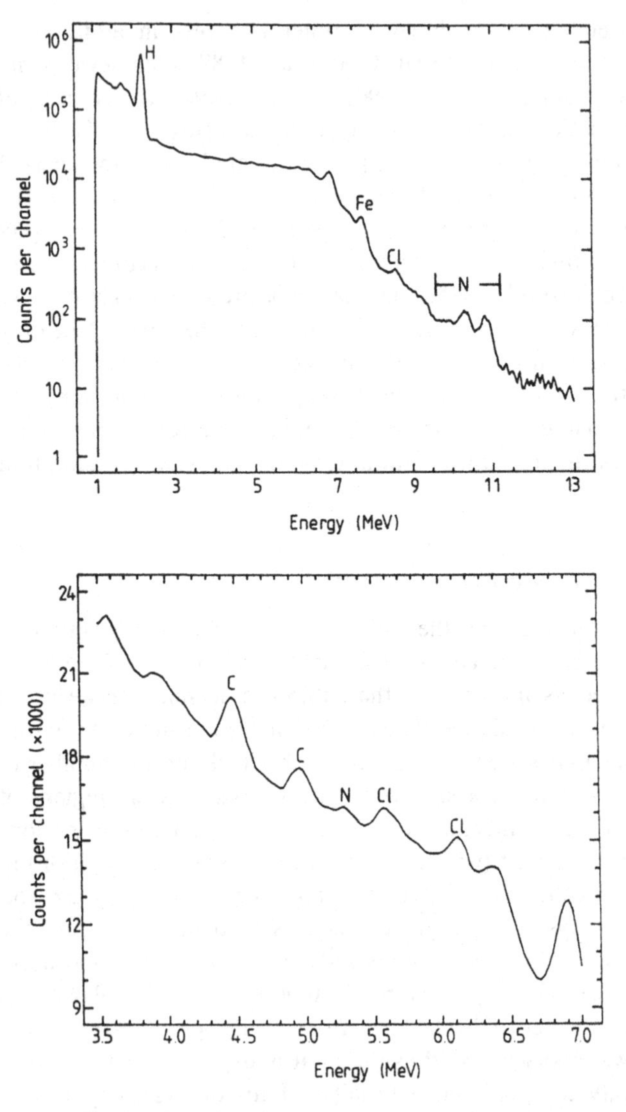

Fig. 2.4. A portion of the prompt *in vivo* gamma-ray spectrum (NaI(Tl) detector) from a shoulder-to-knee bilateral irradiation of a normal male volunteer (dose equivalent of <0.4 mSv). The lower spectrum is an expanded display of the central gamma-ray energy region of the main spectrum. Gamma-ray peaks from carbon, chlorine and nitrogen are visible (but note that there are interferences in these peaks).

by the performance of the gamma-ray spectroscopy apparatus (as is the case during PGNAA) and thus higher dose-rates can be used. This is of benefit to ensure that the decay of activated nuclides is minimised prior to reaching the whole-body counter. The procedure used at Brookhaven

Table 2.2. *Measurement precision, expressed as a coefficient of variation, for various total-body elements measured by IVNAA during recent years*

Element	Reaction	Dose equivalent (mSv)	Precision (%)	Reference[b]
N	$^{14}N(n,\gamma)^{15}N$	<0.5	2–5	1–8
C	$^{12}C(n,n'\gamma)^{12}C$	0.16	3[a]	9
		2	2.5–2.9[a]	10
		<0.5	8–15	2, 11
Cl	$^{35}Cl(n,\gamma)^{36}Cl$	0.45	5.1	2
		0.3	4.9	12
	$^{37}Cl(n,\gamma)^{38}Cl$	<3	2.5	5
Ca	$^{48}Ca(n,\gamma)^{49}Ca$	<3	0.8	5
	$^{40}Ca(n,\gamma)^{41}C$	6.4	2.6	13
Na	$^{23}Na(n,\gamma)^{24}Na$	<3	2.0	5
P	$^{31}P(n,\alpha)^{28}Al$	<3	2.6	5

[a] Obtained using a 14 MeV neutron source; all other measurements were obtained using a radioisotopic neutron source.
[b] References: 1, Beddoe *et al.* (1984*b*); 2, Ryde *et al.* (1989); 3, Krishnan *et al.* (1990); 4, Mackie *et al.* (1990); 5, Dilmanian *et al.* (1990); 6, Stroud *et al.* (1990); 7, Tolli *et al.* (1990); 8, Baur *et al.* (1991); 9, Kehayias *et al.* (1987); 10, Kyere *et al.* (1982); 11, Sutcliffe *et al.* (1990); 12, Mitra *et al.* (1993); 13, Ryde *et al.* (1987).

Laboratory, New York, for example, activates the subject for 5 min followed by 3 min transfer and 15 min counting periods (Dilmanian *et al.*, 1990).

Precision and accuracy

The precision and accuracy of IVNAA are clearly important if measurements are to be of clinical utility (Table 2.2). The actual values of precision and accuracy required for clinical use depend upon the element being measured and its role, for example, in diagnosis.

In general the measurement precision is inversely related to the radiation dose and the gamma-ray detection efficiency. Clearly it is desirable to increase the detection efficiency to obtain the required precision whilst reducing, if possible, the radiation dose. Several other factors, such as the design of the IVNAA instrument, detector shielding and neutron source type, also influence the precision attainable. The precision is normally determined from repeated scans of a tissue-equivalent anthropomorphic phantom under identical conditions to those used for the *in vivo* measurements and expressed as a coefficient of variation. Of all the measurement precisions given in Table 2.2, only those associated with the measurement of carbon using a radioisotope neutron source probably render the measurement too imprecise for clinical use.

For longitudinal or sequential studies in the same subject whose body does not change in size and shape it is the precision that is of paramount importance; for absolute measurements the accuracy also becomes important. Accuracy of IVNAA has been assessed by measuring anthropomorphic phantoms of known composition, by comparison with values obtained for normal subjects by other means or at other centres, and (rarely) by cadaver analysis. The most complete analysis of accuracy using a range of inhomogeneous anthropomorphic phantoms (weight range from 41.4 to 110 kg) at different IVNAA centres was undertaken by Bewley (1988). The results of this study highlighted considerable percentage discrepancy (measured − actual value) for some elements ranging from − 12.3 to 4.8% for calcium, from − 9.7 to 14.6% for nitrogen, and from − 21.2 to 51.0% for chlorine, where all measurements were obtained using DGNAA. Measurement of nitrogen in the same phantoms using PGNAA has provided a percentage discrepancy range from 0.7 to 7.0% (Ryde *et al.*, 1989).

Comparison of total-body nitrogen values obtained in normal subjects with predicted or expected values calculated from regression equations has been pursued by some authors for the assessment of comparability of results and, in general, mean ratios of the measured-to-predicted values have been within 2% of unity with a similar standard error of the mean (Beddoe *et al.*, 1984*a*; Ryde *et al.*, 1993). Comparisons have also been made using various values derived from IVNAA measurements, such as the ratio of total-body nitrogen to the fat-free mass (where, for example, the latter has been derived from a body composition model comprising the sum of water + protein + minerals + glycogen). This type of comparison has shown broad agreement between IVNAA centres and between techniques (Beddoe & Hill, 1985; Mackie *et al.*, 1990; Ryde *et al.*, 1993). In a somewhat similar manner, values of total-body chlorine obtained by IVNAA have been compared with predicted chlorine values and extracellular fluid volume (Beddoe *et al.*, 1987; Ryde *et al.*, 1990*a*; Mitra *et al.*, 1993).

The most robust evaluation of IVNAA accuracy has been obtained from cadvers analysed by both IVNAA and chemical means. This work, most recently undertaken by Knight and colleagues (1986), reported measurement differences between PGNAA and chemical analysis of less than 3% for both total-body nitrogen and chlorine in a 58.6 kg cadaver. Homogenised cadavers were also analysed and the effects of elemental inhomogeneity found to be insignificant.

Recent and future developments

A number of developments have occurred over the past 3–4 years and some of these are summarised here. In general these developments have been directed towards improving the precision of existing measurement techniques by further optimisation of neutron sources, irradiation and counting geometry, gamma-ray detectors and spectroscopy systems. Several of these developments have been directed towards the measurement of total-body nitrogen. Other new techniques such as the resonant scattering of gamma-rays, also for the measurement of nitrogen, are outside the scope of this chapter and will not be considered (Ettinger & Brondo, 1990).

To date, nearly all the IVNAA instruments available in body composition centres throughout the world have been built by local research groups and thus each instrument tends to have a unique fabrication, although there are certain common elements to each design. Whilst this local design and development has without doubt expedited the technical progress of IVNAA, it may also have restricted the clinical use of the technique to those specialist centres where the medical physics expertise for design, fabrication, calibration, quality assurance, routine operation and maintenance is available. This situation has been eased by the recent availability of a commercial instrument, the Body Protein Monitor (Canberra, 1992), based on the design and experience of the Australian Nuclear Science and Technology Organisation. The service, support and training available from a large nuclear instrumentation supplier suggests that such an instrument might be installed without recourse to the broad medical physics expertise noted above, although a professionally qualified medical physicist would still be required to oversee day-to-day operation of the instrument.

The Body Protein Monitor utilises a bilateral irradiation geometry with four NaI(Tl) ($102 \times 102 \times 204$ mm) detectors in a bilateral paired configuration and has a choice of neutron sources (^{252}Cf, ^{241}Am,Be or ^{238}Pu,Be). The instrument has been designed for ease of use and includes menu-driven software control of the scanning table and data acquisition. The subject can be scanned and a report of the result produced within 10 min. For an effective dose equivalent of 0.2 mSv (adults) the accuracy and precision are 4% (Canberra, 1992).

Krishnan and colleagues (1990) have reported a new instrument for total-body nitrogen measurement, designed for use in the hospital environment. In particular, a modified gamma-ray spectroscopy system enhances the gamma-ray signal-to-background ratio in the nitrogen region of interest. The modification exploits the greater abundance ($\sim \times 1000$) of the hydrogen gamma-ray signal compared with that of nitrogen. Adequate counting statistics, therefore, can be achieved for the former in a relatively short time compared with the latter. In practice, the 0–5 MeV gamma-ray energy range (containing the hydrogen gamma-ray at 2.223 MeV) is acquired for 100 s. The 0–5 MeV energy range is then electronically prevented from passing through the spectroscopy system whilst the energy range above 5 MeV is acquired for 1000 s. The effect of this modification is to increase the nitrogen signal-to-background ratio by about 18%.

Studies based upon neutron transport calculations (performed by Monte Carlo and discrete-ordinate computer codes) have been reported recently for the design of a new instrument to measure total-body nitrogen (Stamatelatos *et al.*, 1992). The rationale behind the study was to combine the long half-life advantage of an (α,n) neutron source with the high peak thermal-neutron flux per unit dose advantage of the ^{252}Cf neutron source. This was achieved in the design by fabricating the neutron-source reflector, and the collimator and subject reflector from beryllium and graphite, respectively, to soften the neutron energy spectrum and minimise the leakage of neutrons from the beam collimator and from around the subject. Using a ^{238}Pu,Be neutron source and a laboratory mock-up of the instrument, an improvement in the nitrogen measurement statistics per given dose was reported over existing instruments containing this type of source. Other studies utilising neutron transport codes have also been undertaken (McGregor & Allen 1987; Franklin *et al.*, 1990).

The choice of gamma-ray detectors for the determination of nitrogen has also been examined recently (Stamatelatos *et al.*, 1993). To date, large-volume NaI(Tl) scintillation crystals have been commonly used (typically 152 mm diameter × 152 mm long). However, Stamatelatos and colleagues have shown that five small 51 × 102 mm NaI(Tl) detectors can achieve a similar relative error of nitrogen measurement as one large 152 × 152 mm detector. This arises because small detectors can be more efficiently shielded and the problems of gamma-ray pulse pile-up and gain instability normally associated with high counting rates are reduced.

Other types of detector have also been investigated. In particular, because of its superior efficiency compared with NaI(Tl), the bismuth germanate (BGO) scintillation crystal has been evaluated for the measurement of nitrogen, carbon and chlorine gamma-rays. Whilst the use of these detectors appears to be unsuccessful for the measurement of nitrogen because of interferences in the nitrogen gamma-ray region of interest (arising from activation of germanium in the detector), they are found to be of benefit for the measurement of chlorine and carbon (Blagojevic *et al.*, 1990; Kehayias & Zhuang, 1993).

Pulsed sources of neutrons, for example cyclotrons, were amongst the first sources to be used for the measurement of nitrogen by PGNAA (Biggin *et al.*, 1972). The potential of pulsed sources has continued to be of interest for a wider range of elements (Zamenhof *et al.*, 1979; Kacperek *et al.*, 1990) and has been re-examined recently by Mitra *et al.* (1990) for the measurement, in a single scan, of the major elements carbon, nitrogen, oxygen, chlorine and phosphorus. The objective of this work was the differentiation, with time, of the gamma-ray emissions from the elements into separate spectra according to the type of nuclear reaction and mode of decay. Utilising the beam from a 14 MeV D,T neutron source with pulse length and repetition frequency of 10 μs and 1 kHz, respectively, Mitra *et al.* (1990) proposed a three-phase counting system comprising a neutron 'beam-on' gamma-ray counting period of 10 μs followed by two consecutive 'beam-off' counting periods of about 200 and 800 μs. The initial 10 μs period acquires the prompt gamma-rays arising from fast-neutron inelastic scattering reactions with carbon and oxygen. For the 200 μs immediately following the neutron pulse, the prompt gamma-ray emission arising from thermal-neutron capture is acquired. In the remaining 800 μs it is proposed that delayed gamma-rays are measured, principally from oxygen (via $^{16}O(n,p)^{16}N$), chlorine, calcium and aluminium.

The potential benefit that may be obtained from a pulsed source is illustrated by the work of Sutcliffe *et al.* (1991) reporting the measurement of carbon, hydrogen and oxygen in phantoms with a precision of 9.3%, 2.0% and 8.3%, respectively, for a collimated neutron dose equivalent of 0.17 mSv. Whilst this precision may appear to be of limited clinical use, it should be noted that if the radiation dose of 0.17 mSv were to be distributed over a 1 m section of the body then the dose average would be reduced to less than 10 μSv.

Amongst the most significant developments in IVNAA is a new method for the measurement of carbon and oxygen that enables the magnitude of the gamma-ray background continuum to be reduced and thereby the measurement precision to be increased (Hollas *et al.*, 1990). The method is based on accelerating deuterons onto a tritium target producing 14.7 MeV neutrons and 3.7 MeV alpha particles. The alpha particles and neutrons are produced simultaneously and emitted in opposite directions. The alpha particle is detected and allows a cone beam of neutrons to be defined. Electronic gating of the spectroscopy system enables gamma-rays arising from neutron interactions outside the cone beam to be greatly reduced. Furthermore, time-of-flight analysis permits the point of neutron interaction within the cone to be determined. Hollas *et al.* (1990) report that this arrangement reduces the gamma-ray background by at least 1000 times compared with other pulsed neutron-beam methods and yields precision of better than 3.0% for carbon and oxygen in a 70 kg subject assuming a radiation dose equivalent of 50 μSv. These results were obtained with a single NaI(Tl) scintillation crystal (a cube of dimensions 102 mm). If this method can be successfully applied to *in vivo* measurements then a considerable potential exists for future body composition studies utilising IVNAA.

Conclusions

The development of IVNAA has facilitated a new era in the assessment of the elemental composition of the human body in health and disease. The technique provides unique information for a modest radiation dose. The feasibility of measuring a wide range of elements has been investigated over the last two decades but the most widely available and clinically useful measurement appears to be that of total-body nitrogen. Future applications of the technique are likely to require the continued measurement of nitrogen and possibly carbon, although other elements such as chlorine, phosphorus, sodium, calcium and cadmium may have a minor application. It is not anticipated, however, that the technique will proliferate widely; it is likely to remain confined to a few specialised research centres within the hospital environment. In these centres the technique is expected to make a fundamental and unique contribution to the development and assessment of body composition models, to the evaluation of other (potentially less robust) techniques of body composition, to the compilation of reference data, and to the diagnosis and monitoring of disease.

References

Allen, B.J., Blagojevic, N., McGregor, B.J. *et al.*, (1987). *In vivo* determination of protein in malnourished patients. In *In Vivo Body Composition Studies*, ed. K.J. Ellis, S. Yasumura & W.D. Morgan, pp. 77–83. York: Institute of Physical Sciences in Medicine.

Anderson, J., Osborn, S.B., Tomlinson, R.W.S. *et al.*, (1964). Neutron-activation analysis in man *in vivo*: a new technique in medical investigation. *The Lancet*, **ii**, 1201–5.

Baur, L.A., Allen, B.J., Rose, A., Blagojevic, N. & Gaskin, K.J. (1991). A total-body nitrogen facility for paediatric use. *Physics in Medicine and Biology*, **36**, 1363–75.

Beddoe, A.H. & Hill, G.L. (1985). Clinical measurement of body composition using *in vivo* neutron activation analysis. *Journal of Parenteral and Enteral Nutrition*, **9**, 504–20.

Beddoe, A.H., Streat, S.J. & Hill, G.L. (1984*a*). Evaluation of an *in vivo* prompt gamma neutron activation facility for body composition studies in critically ill intensive care patients: Results on 41 normals. *Metabolism*, **33**, 270–280.

Beddoe, A.H., Zuidmeer, H. & Hill, G.L. (1984*b*). A prompt gamma *in vivo* neutron activation analysis facility for measurement of total body nitrogen in the critically ill. *Physics in Medicine and Biology*, **29**, 371–83.

Beddoe, A.H., Streat, S.J. & Hill, G.L. (1987). Measurement of total body chlorine by prompt gamma *in vivo* neutron activation analysis. *Physics in Medicine and Biology*, **32**, 191–201.

Bewley, D.K. (1988). Anthropomorphic models for checking the calibration of whole-body counters and activation analysis systems. *Physics in Medicine and Biology*, **33**, 805–13.

Biggin, H.C., Chen, N.S., Ettinger, K.V. *et al.*, (1972). Determination of nitrogen in living patients. *Nature New Biology*, **236**, 187–8.

Blagojevic, N., Allen, B.J. & Rose, A. (1990). Development of a total-body chlorine analyser using a bismuth germante detector system and a ^{252}Cf neutron source. In *In Vivo Body Composition Studies*, ed. S. Yasumura, J.E. Harrison, K.G. McNeill, A.D. Woodhead & F.A. Dilmanian, pp. 401–09. New York: Plenum.

Boddy, K. & Alexander, W.D. (1967). Clinical experience of *in vivo* activation analysis of iodine in the thyroid gland – an assesment of the problems. In *Nuclear Activation Techniques in the Life Sciences* 1967, pp. 583–594. Vienna: IAEA.

Boddy, K., Roberston, I. & Glaros, D. (1974). The development of a facility for partial-body *in vivo* activation analysis using ^{252}Cf neutron sources. *Physics in Medicine and Biology*, **19**, 853–61.

Canberra (1992). Body Protein Monitor specification sheet. Canberra Industries, Inc., One State Street, Meriden, Connecticut 06450, USA.

Chamberlain, M.J., Fremlin, J.H., Peters, D.K. & Philips, H. (1968). Total-body calcium by whole-body neutron activation: a new technique for study of bone disease. *British Medical Journal*, **ii**, 581–3.

Chettle, D.R. & Fremlin, J.H. (1984). Techniques of *in vivo* neutron activation analysis. *Physics in Medicine and Biology*, **29**, 1011–43.

Cohn, S.H. (1980). The present state of *in vivo* neutron activation analysis in clinical diagnosis and therapy. *Atomic Energy Review*, **18**, 599–660.

Cohn, S.H. (1981). *In vivo* neutron activation analysis: State of the art and future prospects. *Medical Physics*, **8**, 145–54.

Cohn, S.H. & Dombrowski, C.S. (1971). Measurement of total-body calcium, sodium, chlorine, nitrogen and phosphorus in man by *in vivo* neutron activation analysis. *Journal of Nuclear Medicine*, **12**, 499–505.

Cohn, S.H. & Parr, R.M. (1985). Nuclear-based techniques for the *in vivo* study of human body composition. *Clinical Physics and Physiological Measurement*, **6**, 275–301.

Cohn, S.H., Shukla, K.K., Dombrowski, C.S. & Fairchild, R.G. (1972). Design and calibration of a 'broad-beam' ^{238}Pu,Be neutron source for total-body neutron activation analysis. *Journal of Nuclear Medicine*, **13**, 487–92.

Cohn, S.H., Fairchild, R.G. & Shukla, K.K. (1973). Theoretical considerations in the selection of neutron sources for total-body neutron activation analysis. *Physics in Medicine and Biology*, **18**, 648–57.

De Soete, D., Gijbels, R. & Hoste, J. (1972). *Neutron Activation Analysis*. Chemical Monographs, 34. Chichester: Wiley.

Dilmanian, F.A., Weber, D.A., Yasumura, S. *et al.* (1990). Performance of the delayed- and prompt-gamma neutron activation systems at Brookhaven National Laboratory. In *In vivo Body Composition Studies*, ed. S. Yasumura, J.E. Harrison, K.G. McNeill, A.D. Woodhead & F.A. Dilmanian, pp. 309–17. New York: Plenum.

Ellis, K.J. & Eastman, J.D. ed. (1993). *Human Body Composition. In Vivo Methods, Models and Assessment*. New York: Plenum.

Ellis, K.J. & Kelleher, S.P. (1987). *In vivo* bone aluminium measurements in patients with renal disease. In *In Vivo Body Composition Studies*, ed. K.J. Ellis, S. Yasumura & W.D. Morgan, pp 464–69. York: Institute of Physical Sciences in Medicine.

Ellis, K.J., Yasumura, S. & Morgan, W.D., ed. (1987). *In Vivo Body Composition Studies*. York: Institute of Physical Sciences in Medicine.

Ettinger, K.V. & Brondo, J. (1990). Nuclear resonance fluorescence for the determination of nitrogen *in vivo* (abstract). In *Abstracts of the Annual Scientific Meeting of the IPSM*, Oxford. York: Institute of Physical Sciences in Medicine.

Evans, C.J., Thomas, D.W., Ryde, S.J.S. & Williams, A.J. (1994). Absolute measurements of total-body calcium using prompt-gamma neutron activation analysis: An internal chlorine standardisation method. *Physiological Measurements*, **15**, 67–77.

Franklin, D.M., Amstrong, R., Chettle, D.R. & Scott, M.C. (1990). An improved *in vivo* neutron activation system for measuring kidney cadmium. *Physics in Medicine and Biology*, **35**, 1397–408.

Glaros, D., Xatzikonstantinou, J., Leodiou, J. & Kalef-Ezra, J. (1987). A partial-body activation analysis technique for the measurement of phosphorus in bone. In *In Vivo Body Composition Studies*, ed. K.J. Ellis, S. Yasumura & W.D. Morgan, pp. 295–300. York: Institute of Physical Sciences in Medicine.

Harrison, J.E., McNeill, K.G., Krishnan, S.S. *et al.* (1990). Clinical studies on osteoporosis. In *In Vivo Body Composition Studies*, ed. S. Yasumura, J.E. Harrison, K.G. McNeill, A.D. Woodhead & F.A. Dilmanian, pp. 83–8. New York: Plenum.

Haywood, J.K., Williams, E.D., McArdle, F.J. & Boddy, K. (1981). Reliability of absolute and relative measurement of total-body nitrogen by the $^{14}N(n, 2n)^{13}N$ reaction. *Physics in Medicine and Biology*, **26**, 591–602.

Hoffman, J.G. & Hempelmann, L.H. (1957). Estimation of whole-body radiation doses in accidental fission bursts. *American Journal of Roentgenology*, **77**, 144–60.

Hollas, C.L., Ussery, L.E., Butterfield, K.B. & Morgado, R.E. (1990). A method for *in vivo* determination of carbon and oxygen using prompt-gamma radiations induced by 14.7 MeV neutrons. In *In Vivo Body Composition Studies*, ed. S. Yasumura, J.E. Harrison, K.G. McNeill, A.D. Woodhead & F.A. Dilmanian, pp. 395–401. New York: Plenum.

Kacperek, A., Morgan, W.D., Sivyer, A., Dutton, J. & Evans, C.J. (1990). The application of a pulsed fast-neutron beam to partial body *in vivo* activation analysis of minerals and trace elements. *Journal of Radioanalytical and Nuclear Chemistry, Articles*, **140**, 141–51.

Kehayias, J.J., Ellis, K.J., Cohn, S.H., Yasumura, S. & Weinlein, J.H. (1987). Use of a pulsed neutron generator for *in vivo* measurement of body carbon. In *In Vivo Body Composition Studies*, ed. K.J. Ellis, S. Yasumura & W.D. Morgan, pp. 427–36. York: Institute of Physical Sciences in Medicine.

Kehayias, J.J., S.B. Heymsfield, S.B., Dilmanian, F.A., Wang. J., Gunther, D.M. & Pierson, R.N. (1990). Meaurement of body fat by neutron inelastic scattering: comments on installation, operation and error analysis. In *In Vivo Body Composition Studies*, ed. S. Yasumura, J.E. Harrison, K.G. McNeill, A.D. Woodhead & F.A. Dilmanian, pp. 339–47. New York: Plenum.

Kehayias, J.J. & Zhuang, H. (1993). Measurement of regional body fat *in vivo* in humans by simultaneous detection of regional carbon and oxygen using neutron inelastic scattering at low radiation exposure. In *Human Body Composition. In Vivo Methods, Models and Assessment*, ed. K.J. Ellis & J.D. Eastman, pp. 49–52. New York: Plenum.

Kennedy, N.S.J., Eastell, R., Ferrington, C.M. *et al.*, (1982). Total-body neutron acivation analysis of calcium: calibration and normalisation. *Physics in Medicine and Biology*, **27**, 697–707.

Kennedy, N.S.J., Eastell, R., Smith, M.A. & Tothill, P. (1983). Normal levels of total-body sodium and chlorine by neutron activation analysis. *Physics in Medicine and Biology*, **28**, 215–21.

Knight, G.S., Beddoe, A.H., Streat, S.J. & Hill, G.L. (1986). Body composition of two human cadavers by neutron activation and chemical analysis. *American Journal of Physiology* (*Endocrinology and Metabolism*, 13), **250**, E179–85.

Knoll, G.F. (1988). *Radiation Detection and Measurement*, 2nd edn. New York: Wiley.

Krishnan, S.S., McNeill, K.G., Mernagh, J.R., Bayley, A.J. & Harrison, J.E. (1990). Improved clinical facility for *in vivo* nitrogen measurement. *Physics in Medicine and Biology*, **35**, 489–499.

Kyere, K., Oldroyd, B., Oxby, C.B., Burkinshaw, L., Ellis, R.E. & Hill, G.L. (1982). The feasibility of measuring total-body carbon by counting neutron inelastic scatter gamma rays. *Physics in Medicine and Biology*, **27**, 805–17.

Larsson, L., Alpsten, M., Tolli, J., Drugge, N. & Mattsson, S. (1987). *In vivo* measurements of nitrogen using a neutron activation technique. In *In Vivo Body Composition Studies*, ed. K.J. Ellis, S. Yasumura & W.D. Morgan, pp. 436–40. York: Institute of Physical Sciences in Medicine.

Mackie, A., Hannan, W.J., Smith, M.A. & Tothill, P. (1988). Apparatus for the measurement of total-body nitrogen using prompt neutron activation analysis with ^{252}Cf. *Journal of Medical Engineering and Technology*, **12**, 152–9.

Mackie, A., Cowen, S. & Hannan J. (1990). Calibration of a prompt neutron activation analysis facility for the measurement of total-body protein. *Physics in Medicine and Biology*, **35**, 613–24.

McCarthy, I.D., Sharafi, A., Oxby, C.B., & Burkinshaw, L. (1980). The accuracy of total-body nitrogen determined by neutron activation analysis. *Physics in Medicine and Biology*, **25**, 849–63.

McGregor, B.J. & Allen, B.J. (1987). Monte Carlo study of neutron and gamma-ray transport in the AAEC total body nitrogen facility. *Australasian Physical and Engineering Sciences in Medicine*, **10**, 155–61.

McNeill, K.G., Thomas, B.J., Sturtridge, W.C. & Harrison, J.E. (1973). *In vivo* neutron activation analysis for calcium in man. *Journal of Nuclear Medicine*, **14**, 502–6.

McNeill, K.G., Harrison, J.E., Mernagh, J.R., Stewart, S. & Jeejeebhoy, K.N. (1982). Changes in body protein, body potassium, and lean body mass during total parenteral nutrition. *Journal of Parenteral and Enteral Nutrition*, **6**, 106–8.

McNeill, K.G., Borovnicar, D.J., Krishnan, S.S., Wang, H., Wanna, C. & Harrison, J.E. (1989). Investigation of factors which lead to the background in the measurement of nitrogen by IVNAA. *Physics in Medicine and Biology*, **34**, 53–9.

Mernagh, J.R., Harrison, J.E. & McNeill, K.G. (1977). *In vivo* determination of nitrogen using Pu,Be sources. *Physics in Medicine and Biology*, **22**, 831–5.

Mitra, S., Sutcliffe, J.F. & Hill, G.L. (1990). A proposed three-phase counting system for the *in vivo* measurement of the major elements using pulsed 14 MeV neutrons. *Biological Trace Element Research*, **26**–27, 423–8.

Mitra, S., Plank, L.D., Knight, G.S. & Hill, G.L. (1993). *In vivo* measurement of total-body chlorine using the 8.57 MeV prompt de-excitation following thermal neutron capture. *Physics in Medicine and Biology*, **38**, 161–172.

Morgan, W.D., Vartsky, D., Ellis, K.J. & Cohn, S.H. (1981). A comparison of ^{252}Cf and ^{238}Pu,Be neutron sources for partial-body *in vivo* activation analysis. *Physics in Medicine and Biology*, **26**, 413–24.

Morgan, W.D., Ryde, S.J.S., Jones, S.J. *et al.* (1990). *In vivo* measurements of cadmium and lead in occupationally exposed workers and an urban population. *Biological Trace Element Research*, **26**–27, 407–14.

Mountford, P.J. (1982). Effect of collimator and depth in a phantom on the neutron spectrum from a ^{238}Pu,Be neutron source. *Physics in Medicine and Biology*, **27**, 1245–52.

Nelp, W.B., Palmer, H.E., Murano, R. *et al.* (1970). Measurement of total-body calcium (bone mass) *in vivo* with the use of total-body neutron activation analysis. *Journal of Laboratory Clinical Medicine*, **76**, 151–62.

Ryde, S.J.S., Morgan, W.D., Sivyer, A., Evans, C.J. & Dutton, J. (1987). A clinical instrument for multi-element *in vivo* analysis by prompt, delayed and cyclic neutron activation using ^{252}Cf. *Physics in Medicine and Biology*, **32**, 1257–71.

Ryde, S.J.S., Morgan, W.D., Evans, C.J., Sivyer, A. & Dutton J. (1989). Calibration and evaluation of a ^{252}Cf-based neutron activation analysis instrument for the determination of nitrogen *in vivo*. *Physics in Medicine and Biology*, **34**, 1429–41.

Ryde, S.J.S., Morgan, W.D., Thomas, D.W. *et al.* (1990*a*). Prompt-gamma measurements of nitrogen and chlorine in normal volunteers. In *In Vivo Body Composition Studies*, ed. S. Yasumura, J.E. Harrison, K.G. McNeill, A.D. Woodhead & F.A. Dilmanian, pp. 347–53. New York: Plenum.

Ryde, S.J.S., Morgan, W.D., Compston, J.E., Evans, C.J., Sivyer, A. & Dutton, J. (1990*b*). Measurement of total-body calcium by prompt-gamma neutron activation analysis using a ^{252}Cf source. *Biological Trace Element Research*, **26**–27, 429–438.

Ryde, S.J.S., Morgan, W.D., Birks, J.L., Evans, C.J. & Dutton, J. (1993). A five-compartment model of body composition of healthy subjects assessed using *in vivo* neutron activation analysis. *European Journal of Clinical Nutrition*, **47**, 863–74.

Ryde, S.J.S., Bowen-Simpkins, K., Bowen-Simpkins, P., Evans, W.D. *et al.* (1994). The effect of oestradiol implants on regional and total bone mass: a three year longitudinal study. *Clinical Endocrinology*, **40**, 33–8.

Scott, M.C. & Chettle, D.R. (1986). *In vivo* elemental analysis in occupational medicine. *Scandinavian Journal of Work in Environmental Health*, **12**, 81–96.

Sharafi, A., Pearson, D., Oxby, C.B., Oldroyd, B., Krupowicz, D.W., Brooks, K. & Ellis, R.E. (1983). Multi-element analysis of the human body using neutron activation. *Physics in Medicine and Biology*, **28**, 203–14.

Spinks, T.J., Bewley, D.K., Ranicar, A.S.O. & Joplin, G.F. (1977). Measurement of total-body calcium in bone disease. *Journal of Radioanalytical Chemistry*, **37**, 345–55.

Spinks, T.J., Bewley, D.K., Paolillo, M., Vlotides, J., Joplin, G.F. & Ranicar, A.S.O. (1980). Metabolic activity of sodium, measured by neutron activation, in the hands of patients suffering from bone diseases: concise communication. *Journal of Nuclear Medicine*, **21**, 41–6.

Stamatelatos, I.E.M., Chettle, D.R., Green, S. & Scott, M.C. (1992). Design studies related to an *in vivo* neutron activation analysis facility for measuring total-body nitrogen. *Physics in Medicine and Biology*, **37**, 1657–74.

Stamatelatos, I.E.M., Chettle, D.R. & Scott, M.C. (1993). Studies relating to the choice of gamma-ray detectors for *in vivo* nitrogen measurement by prompt-capture neutron activation analysis. *Physics in Medicine and Biology*, **38**, 411–22.

Stroud, D.B., Borovnicar, D.J., Lambert, J.R. *et al.* (1990). Clinical studies of total-body nitrogen in an Australian Hospital. In *In Vivo Body Composition Studies*, ed. S. Yasumura, J.E. Harrison, K.G. McNeill, A.D. Woodhead & F.A. Dilmanian, pp. 177–83. New York: Plenum.

Sutcliffe, J.F., Mitra, S. & Hill, G.L. (1990). *In vivo* measurement of total-body carbon using ^{238}Pu,Be neutron sources. *Physics in Medicine and Biology*, **35**, 1089–98.

Sutcliffe, J.F., Waker, A.J., Smith, A.H., Barker, M.C. & Smith, M.A. (1991). A feasibility study for the simultaneous measurement of carbon, hydrogen and oxygen using pulsed 14.4 MeV neutrons. *Physics in Medicine and Biology*, **36**, 87–98.

Tolli, J., Alpsten, M., Larsson, L. *et al.* (1990). Total-body nitrogen and potassium determination in patients during Cis-Pt treatment. In *In Vivo Body Composition Studies*, ed. S. Yasumura, J.E. Harrison, K.G. McNeill, A.D. Woodhead & F.A. Dilmanian, pp. 171–76. New York: Plenum.

Vartsky, D., Ellis, K.J. & Cohn, S.H. (1979*a*). *In vivo* measurement of body nitrogen by analysis of prompt-gammas from neutron capture. *Journal of Nuclear Medicine*, **20**, 1158–65.

Vartsky, D., Prestwich, W.V., Thomas, B.J. *et al.* (1979*b*). The use of body hydrogen as an internal standard in the measurement of nitrogen *in vivo* by prompt neutron capture gamma-ray analysis. *Journal of Radioanalytical Chemistry*, **48**, 243–52.

Vartsky, D., Ellis, E.J., Vaswani, A.N., Yasumura, S. & Cohn, S.H. (1984). An improved calibration of the *in vivo* determination of body nitrogen, hydrogen and fat. *Physics in Medicine and Biology*, **29**, 209–18.

Williams, E.D., Boddy, K., Harvey, R. & Haywood, J.K. (1978). Calibration and evaluation of a system for total-body *in vivo* activation analysis using 14 MeV neutrons. *Physics in Medicine and Biology*, **23**, 405–15.

Wyatt, R.M., Ryde, S.J.S., Morgan, W.D., McNeill, E.A., Hainsworth, I.R. & Williams, A.J. (1993). The development of a technique to measure bone aluminium content using neutron activation analysis. *Physiological Measurements*, **14**, 327–35.

Yasumura, S., Harrison, J.E., McNeill, K.G., Woodhead, A.D. & Dilmanian, F.A., ed. (1990). *In Vivo Body Composition Studies*. New York: Plenum.

Zamenhof, R.G., Deutsch, O.L. & Murray, B.W. (1979). A feasibility study of prompt capture gamma *in vivo* neutron activation analayis. *Medical Physics*, **6**, 179–92.

3 *Magnetic resonance imaging for the assessment of body composition*

PAOLO BRAMBILLA, PAOLA MANZONI, PAOLO SIMONE AND GIUSEPPE CHIUMELLO

Introduction

Body composition assessment has become of great importance in evaluating physiological and pathological states. In obesity research estimation of total-body fat and fat regional distribution are considered an important goal both in adults and in children. In fact, the morbidity risk of obesity has been clearly related to the amount of total-body fat and, more strongly, to visceral adiposity (Krotkiewsky *et al.*, 1983,; Peiris *et al.*, 1989). Consequently the validation of different techniques for body composition analysis has become of great importance. New methods have been added to old ones, but it is still hard to define the optimum in fulfilling all the necessary criteria. This is particularly true for children, where ethics, accuracy and acceptable cost are fundamental. In Table 3.1 some of the available techniques are rooted according to these characteristics: none of them fully satisfies our criteria and only a few can discriminate body fat distribution.

Magnetic resonance imaging (MRI) has recently been proposed for assessing fat content and distribution in both adults and children. This technique may supersede computed tomography (CT) scanning and therefore avoid ionising radiation exposure. Table 3.1 identifies MRI as ideal in terms of ethics and precision, but its high cost limits its use to selected populations.

MRI technique is based on the interaction between the nuclei of hydrogen atoms, which occur abundantly in all biological tissues, and the magnetic field, 10 000 times stronger of that of the Earth, generated by the MRI system. When a subject is placed inside the field, protons tend to align themselves with the field. A radiofrequency pulse is then applied to the system causing an absorption of energy. After turning off the radiofrequency pulse, protons release this energy in the form of a radiofrequency signal, used by a computer to develop the MR images (Foster *et al.*, 1984). The spin-echo data acquisition technique increases the different T1 relaxation times of adipose tissue and muscle to provide MR images of high quality and definition (Dooms *et al.*, 1986; Foster *et al.*, 1984).

Table 3.1. *Applicability score in paediatric studies for some of the available techniques for body composition analysis*

Technique	Ethical rating	Low cost	Accuracy for fat estimation	Precision for fat distribution
Anthropometry	3	3	1	2
Bioelectrical impedance	3	2	1	0
MRI	3	0	3	3
Deuterium oxide dilution	3	1	1	0
DXA	2	0	3	2
TOBEC	3	0	1	0

Derived from Lukaski (1987).
MRI, magnetic resonance imaging; DXA, dual-energy X-ray absorptiometry; TOBEC, total body electrical conductivity.
0, none; 1, poor; 2, good; 3, optimal.

In comparison with CT, MRI is preferable as regards the absence of ionising radiation, its relevance in paediatric studies, and the absence of interference by bone tissue in the signal from fat and muscle. However, CT is considered more accurate for imaging visceral fat and has a superior capability for distinguishing between different visceral fat compartments (Seidell *et al.*, 1990).

MRI studies of adults

Numerous studies have demonstrated that MRI can accurately measure adipose tissue in animals and both subcutaneous and intra-abdominal adipose areas in humans.

Ross *et al.* (1992) suggested the lumbar region as the most representative for both subcutaneous and visceral adipose tissue, after the analysis of 41 slices from head to toe in healthy adults.

Seidell *et al.* (1990) compared CT and MRI scan results at the lumbar level in healthy adults. The agreement between the two imaging techniques was satisfactory for total adipose tissue at the lumbar level as well for subcutaneous adiposity. However, MRI underestimated the visceral adipose tissue area, even though CT and MRI measures remained significantly correlated. The reasons for this discrepancy are many. First, there are relatively more pixels in visceral fat and visceral adipose tissue borders not only muscle but also the irregular boundaries of the intestinal wall. Furthermore intestinal movements can cause artefacts which are enhanced by the longer acquisition time for MRI in comparison with CT.

The comparison of MRI with a non-imaging technique, such as underwater weighing (UWW), gave less than perfect results (Fowler *et al.*, 1991). This is to be expected as UWW estimates the relative proportions of fat and fat-free mass rather than adipose tissue directly.

The clinical importance of the amount of visceral fat is well documented and outlined by a MRI study in obese women (Gray *et al.*, 1991). A larger abdominal visceral fat content was found in subjects with non-insulin dependent diabetes than in glucose-tolerant women. After adequate weight loss on a very low energy diet a significant decrease in both subcutaneous and visceral adipose tissue was observed. MRI studies seem to offer specific support in clinical monitoring of body fat variations during diet programmes.

MRI studies of children

The impact of pubertal development on fat distribution has been evaluated in normal girls (De Ridder *et al.*, 1992). MRI analysis was carried out at three different levels corresponding to the sites of anthropometric circumference measurements: waist, hip and trochanter. Early and late pubertal girls differed only as regards the trochanteric adipose areas, which were higher in the older group. The range of intra-abdominal fat areas was smaller in comparison with adult female values, suggesting that in late female puberty adipose tissue is predominantly stored in the gluteal and not in the abdominal area.

In a group of children equally distributed in all the body mass index (BMI) quintiles, Fox *et al.* (1993) found a lower adipose abdominal content in comparison with adults and a lack of correlation between visceral adipose tissue as detected by MRI and the waist/hip ratio.

We have undertaken studies in 23 (11 boys, 12 girls) obese and 21 (12 boys, 9 girls) normal children aged 10–15 years. Mean BMI was 29.6 $\pm$ 3.8 in the obese group and 17.4 $\pm$ 1.9 in the controls. A Toshiba 50 A-MRT machine was used, with a spin-echo sequence, T1-weighted relaxation time, and 300-20 acquisition time. The same trained operator performed all the scans and subsequent analysis. Adipose tissue areas were defined using a manual trackball technique. Computed calculation of each area was then performed. Each subject underwent scans at two levels: in the arm, at the mid-humerus level (scan time 5 min, analysis time 10 min), and in the abdomen at the lumbar (L4) level, corresponding to the umbilicus (scan time 10 min, analysis time 20 min). We computed two ratios: subcutaneous adipose area/total L4 area (S/TA) and intra-abdominal adipose area/intra-abdominal area (I/IA).

At mid-humerus level both execution and analysis time were fast and the scan was easy to perform, with good image resolution for fat (Fig. 3.1).

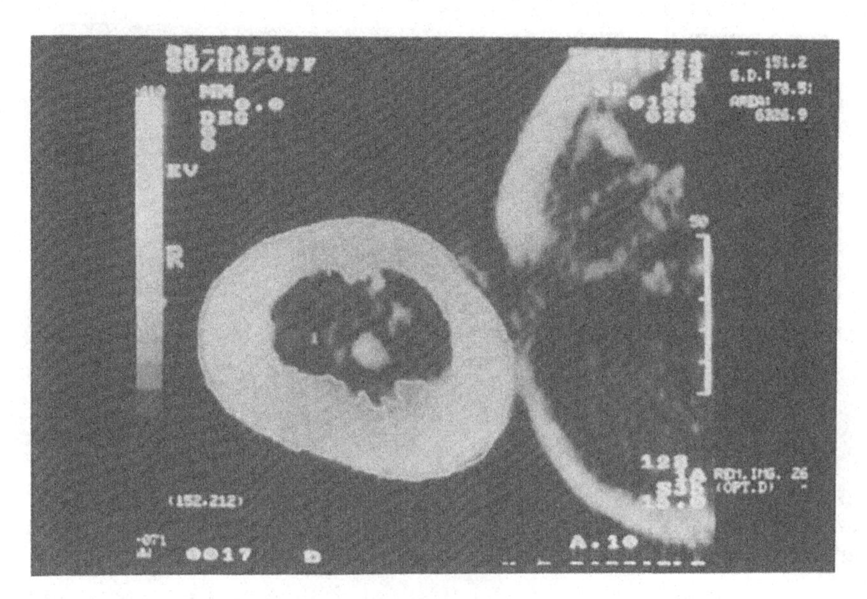

Fig. 3.1. MRI image at the mid-humerus in an obese subject.

Adipose areas and percentage were significantly higher in obese children and highly correlated with arm fat area (AFA) estimated from the triceps skinfold thickness and arm circumference (Sann *et al.*, 1988) ($r = 0.84$, $p < 0.0005$). Using the mean of the biceps and triceps skinfolds instead of triceps only in the formula did not result in a stronger correlation ($r = 0.60$, $p < 0.06$), probably as a consequence of the lower precision of biceps skinfold measurement compared with triceps. At the lumbar level execution time was still short but analysis required more time, especially to identify and calculate visceral adipose areas, while subcutaneous fat showed good image resolution (Fig. 3.2).

The subcutaneous adipose area was highly represented in all obese children, ranging from 55% to 75% of total L4 area, with a relatively narrow range. In normal subjects this ratio was significantly lower (7–57 %) except in two children, who were defined as normal according to anthropometry. The overlap in S/TA ranges in obese subjects and controls highlights the imprecision of the standard for defining childhood obesity. This ratio was positively correlated with some anthropometric indexes in obese children, suggesting a clinical significance as an index of total-body adiposity. Obese subjects showed higher intra-abdominal fat content (I/IA $20 \pm 6\%$) than controls ($16 \pm 13\%$), even though widespread variation was observed in both groups. Intra-abdominal fat was, however, markedly lower in comparison with previous data from adults with central obesity.

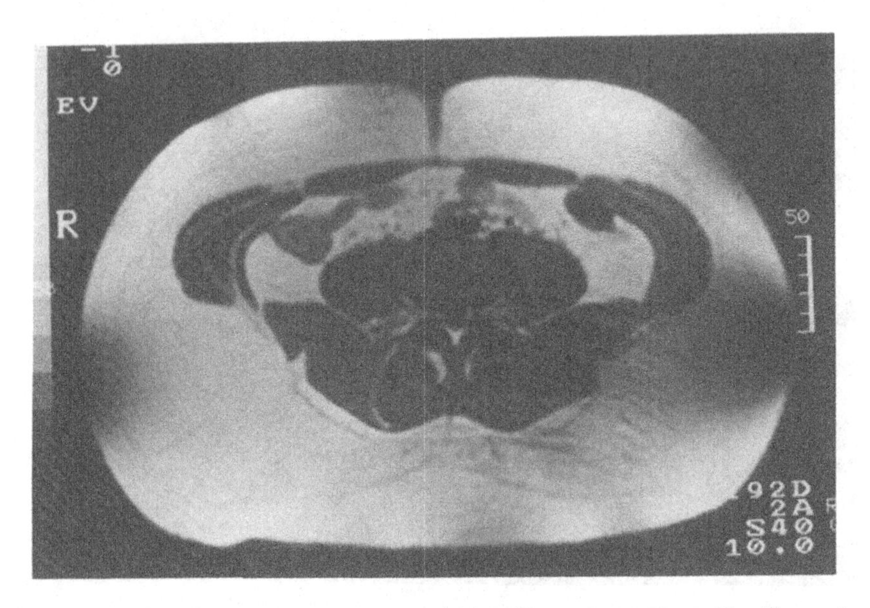

Fig. 3.2. MRI image at the lumbar level (L4) in an obese subject, with evidence of subcutaneous adiposity and intra-abdominal fat content.

No differences were found regarding sex and pubertal status in the obese group.

No correlations were found with adiposity indexes but I/IA was significantly correlated with total blood cholesterol level ($r = 0.57, p < 0.01$) and low density lipoprotein (LDL) fraction ($r = 0.60$, $p < 0.006$), as well with insulin response to oral glucose load at 120 min ($r = 0.50$, $p < 0.03$). These data seem to confirm the important role of visceral adiposity in metabolic complications, even in children. Fig. 3.3 plots visceral adiposity versus waist/hip ratio in obese children, and shows the absence of correlation between this commonly used index of fat distribution in adults and intra-abdominal adiposity.

Conclusion

MRI is a feasible technique for body composition studies, especially for the analysis of fat distribution. In obese children, in particular, when CT scanning techniques are not applicable for ethical reasons, MRI seems to offer a valid alternative. However, while MRI has many advantages, such as repeatability in monitoring studies (Fuller *et al.*, 1990), its high cost limits its application to selected populations. Our experience in childhood obesity suggests the clinical value of abdominal MRI studies in discriminating

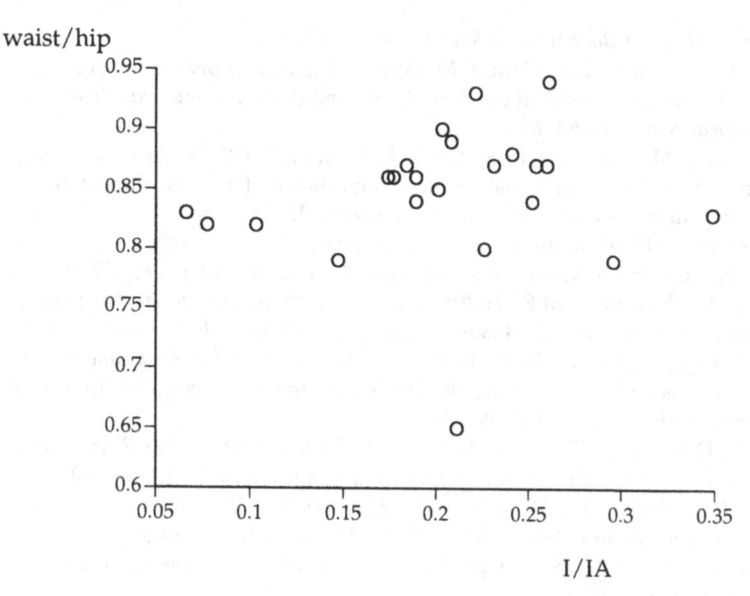

Fig. 3.3. Lack of correlation between visceral adiposity (I/IA) and waist/hip ratio in our study population.

subcutaneous and visceral adiposity. The first can be interpreted as an index of total-body fatness, while the low content of intra-abdominal fat needs follow-up studies for defining thc timing and possible regulatory factors in the development of the adult adiposity pattern.

References

De Ridder, C.M., de Boer, R.W., Seidell, J.C., *et al.* (1992). Body fat distribution in pubertal girls quantified by magnetic resonance imaging. *International Journal of Obesity*, **16**, 443–449.

Dooms, G.C., Hricak, H., Margulis, A.R. & de Geer, G. (1986). MRI imaging of fat. *Radiology*, **158**, 51–54.

Foster, M.A., Hutchison, J.M.S., Mallard, J.R. & Fuller, M.F. (1984). Nuclear magnetic resonance pulse sequence and discrimination of high and low fat tissues. *Magnetic Resonance Imaging*, **2**, 187–192.

Fowler, P.A., Malcolm, F.F. & Glasbey, C.A. (1991). Total and subcutaneous adipose tissue in women: the measurement of distribution and accurate prediction of quantity by using magnetic resonance imaging. *American Journal of Clinical Nutrition*, **54**, 18–25.

Fox, K., Peters, D., Armstrong, N., Sharpe, P. & Bell, M. (1993). Abdominal fat deposition in 11-year-old children. *International Journal of Obesity*, **17**, 11–16.

Fuller, M.F., Fowler, P.A., McNeill, G. & Foster, M.A. (1990). Body composition: the accuracy of new methods and their suitabilty for longitudinal studies.

Proceedings of the Nutrition Society, **49**, 423–36.

Gray, D.S., Fujioka, K., Colletti, P.M. (1991). Magnetic resonance imaging used for determining fat distribution in obesity and diabetes. *American Journal of Clinical Nutrition*, **54**, 623–7.

Krotkiewsky, M., Bjorntorp, P., Sjostrom, L. & Smith, U. (1983). Impact of obesity on metabolism in men and women. Importance of regular adipose tissue distribution. *Journal of Clinical Investigation*, **72**, 1150–62.

Lukaski, H.C. (1987). Methods for the assessment of human body composition: traditional and new. *American Journal of Clinical Nutrition*, **46**, 537–56.

Peiris, A.N., Sothmann, M.S., Hoffman, R.G. (1989). Adiposity, fat distribution and cardiovascular risk. *Annals of Internal Medicine*, **110**, 867–72.

Ross, R., Leger, L., Morris, D., De Guise, J. & Guardo, R. (1992). Quantification of adipose tissue by MRI: relationship with anthropometric variables. *Journal of Applied Physiology*, **72(2)**, 787–95.

Sann, L., Durand, M., Picard, J., Lasne, Y. & Bethenod, M. (1988). Arm fat and muscle areas in infancy. *Archives of Disease in Childhood*, **63**, 256–260.

Seidell, J.C., Bakker, C.J.G. & Van Der Kooy (1990). Imaging techniques for measuring adipose tissue distribution. A comparison between computed tomography and 1.5 T magnetic resonance. *American Journal of Clinical Nutrition*, **51**, 953–7.

4 *Multi-frequency impedance as a measure of body water compartments*

PAUL DEURENBERG

Introduction

Following the early work of Thomasset (1962), Hoffer *et al.* (1969) and Nijboer (1970), and the more recent study of Lukaski *et al.* (1985*b*), the bioelectrical impedance method has become a popular way of assessing body composition. This may be due at least in part to its simplicity. Numerous studies have shown that the methodology provides valid estimates of total body water and fat-free mass (Lukaski *et al.*, 1985; Kushner & Schoeller, 1986; Segal *et al.*, 1988). Prediction formulae are, however, population-specific (Deurenberg *et al.*, 1991), and the accuracy and validity of the methodology is, as some authors have pointed out, not a real improvement over other methods such as anthropometry (Diaz *et al.*, 1989; Deurenberg *et al.*, 1991; Kooy *et al.*, 1992). There is disagreement as to whether the method can predict changes in body composition (Deurenberg *et al.*, 1989; Kushner *et al.*, 1990; de Lorenzo *et al.*, 1991; Vasquez & Janoski, 1991; Forbes *et al.*, 1992). It has been demonstrated that body impedance at 50 kHz depends on the distribution between extra- and intracellular water, body compartments which differ in specific resistivity (Deurenberg *et al.*, 1989). Consequently changes in body water that alter the ratio of extra- to intracellular water are not predicted accurately.

Recently a new generation of impedance instruments has become available. They differ from the first generation in that they are able to measure body impedance at more than one frequency. Modern electronics allow the quick measurement of impedance at a series of frequencies, ranging from low (about 1 kHz) to very high (>1 MHz). There are, however, practical difficulties in measuring body impedance at such extreme frequencies, due for example to contact resistance between the electrodes and the skin, and the characteristics of the current at high frequency.

It is theoretically possible to distinguish between the extra- and intracellular water compartments using multi-frequency impedance (Jenin *et al.*, 1975; Settle *et al.*, 1980). At low frequency the current is unable to pass through the cell membrane due to its high capacitive resistance. Hence, at low frequency body impedance should be a measure of extracellular water only.

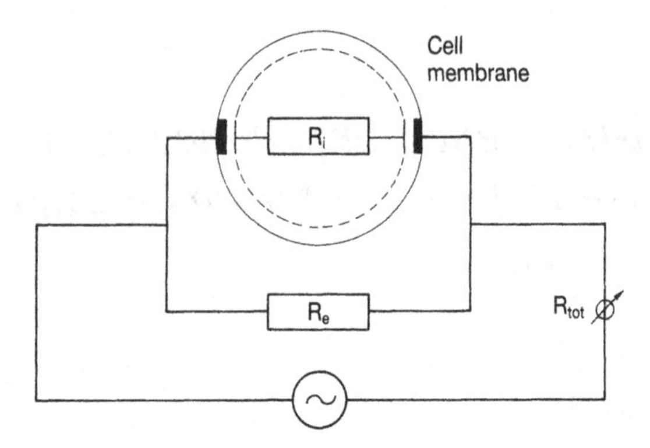

Fig. 4.1. Electrical pathway in the body at different frequencies, assuming a parallel connection of the intra- and extracellular volumes. R_{tot}, total resistance; R_i, intracellular resistance; R_e, extracellular resistance. At low frequency, $R_{tot} = R_e$. At high frequency, $R_{tot} = (R_e \times R_i)/(R_e + R_i)$ and $R_i = (R_e \times R_{tot})/(R_e - R_{tot})$.

With increasing frequency the capacitive resistance of the cell membrane diminishes, so that at high frequency body impedance is a measure of total body water (Fig. 4.1).

This chapter focuses on the assessment of body water compartments by multi-frequency bioelectrical impedance. Prediction formulae for extracellular water and total body water are developed and cross-validated. The formulae are applied in a study where changes in body water and body water compartments occur. It is shown that although body water compartments can be predicted accurately in cross-sectional studies, the same is not true of changes in body water, either at an individual or at a group level. Possible theoretical reasons for this phenomenon are discussed.

Subjects and methods

The study was performed in 103 healthy males and females, aged 19–51 years. Five subjects volunteered for a further study of diuretic-induced water loss, where the aim was to validate the prediction formulae for body water compartments developed for the whole sample. The study protocols were approved by the Medical Ethical Committee of the Department of Human Nutrition, Wageningen Agricultural University.

All measurements were performed in the morning, after an overnight fast and after emptying the bladder. Body weight was measured on a digital scale to the nearest 0.05 kg. Body height was measured by means of a Microtoise to the nearest 1 mm. Four-electrode bioelectrical impedance

(Ω) was measured on the left side of the body, following the methodology of Lukaski *et al.* (1985). In the diuretic study the electrode positions were marked with a marker pencil to improve reliability. Impedance was measured at a large number of frequencies, ranging from 0.3 kHz to 100 kHz, using a Human-Im Scan (Dieto-System, Milan, Italy). Adhesive electrodes with a surface area of 5 cm^2 were used (Littmann 2325VP, 3M, St Paul, MN, USA). From the impedance scans the extra cellular resistance (R_e) and total resistance (R_{tot}) were derived using non-linear regression (see later). Total body water (TBW) was determined by deuterium oxide dilution, and extra-cellular water (ECW) by potassium bromide dilution. The tracer dose (a cocktail of deuterium oxide and potassium bromide) was taken orally immediately before the other measurements. After 2.5 h dilution time a venous blood sample was drawn. After sublimation of the plasma, the deuterium concentration was determined in the sublimate by infrared analysis (Lukaski & Johnson, 1985). TBW was calculated from the given dose and the tracer concentration determined in plasma, using a correction of 5% for non-aqueous dilution (Forbes, 1987). Bromide in plasma was determined after ultra-filtration by high performance liquid chromatography (Miller & Cappon, 1984). A correction of 5% was used for the Donnan effect and a correction of 10% for non-extracellular dilution (Forbes, 1987). Intracellular water (ICW) was calculated as the difference between TBW and ECW.

In the diuretic study body weight was measured before and after taking frusemide (40 mg Lasix, Hoechst, Germany), and the voided urine was collected. Sodium and potassium in urine were determined by atomic absorption spectrophotometry (Perkin-Elmer 2380, Norwalk, CT, USA). The excretion of ECW was calculated from the sodium excretion, assuming an extracellular sodium concentration of 142 mmol/l (Forbes, 1987).

A mathematical model relating impedance (Ω) to frequency (kHz) was fitted to the data for each subject over the frequency range 1 kHz to 100 kHz, using the SPSS/PC-v4.0 program (1990):

$$\text{Impedance} = b - \frac{a(c-b)}{(\text{frequency} - a)}$$

Estimates of extracellular resistance (R_e) and total resistance (R_{tot}) were derived from the constants b and c, corresponding to frequency zero ($R_e = c$) and infinity ($R_{tot} = b$) respectively. The correlation coefficient obtained for the regression model was consistently higher than 0.999, as shown in Fig. 4.2. The estimation errors from the impedance scan (coefficients of variation) of R_e and R_{tot} were calculated to be 1% and 3% respectively. The intra cellular resistance (R_i) was calculated from R_e and

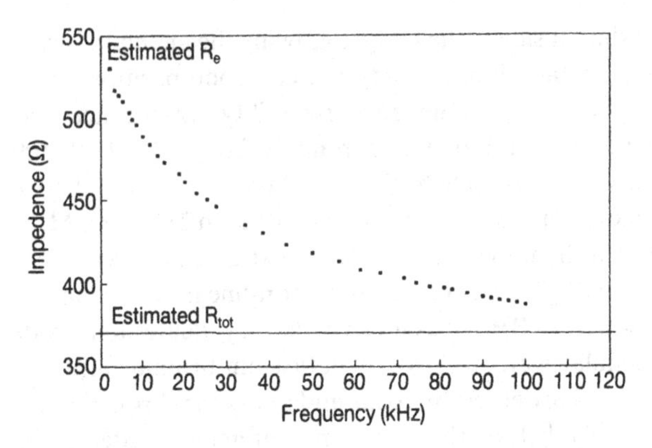

Fig. 4.2. The mathematical model of the relation between impedance and frequency. For details see text.

R_{tot}, assuming that ECW and ICW act as two resistors in parallel ($1/R_i = 1/R_{tot} - 1/R_e$; Fig. 4.1).

Pearson's correlation coefficients and partial correlation coefficients were calculated between TBW and ECW on the one hand and height^2/R_e (H^2/R_e) and H^2/R_{tot} on the other. Prediction formulae were developed using stepwise multiple linear regression. For this purpose the total study population was randomly divided into two subgroups, and the prediction equations as developed in each subgroup were validated in the other. Differences between males and females, and between groups differing in body mass index (weight/height^2) or age, were tested using Student's *t*-test. Values are expressed throughout as mean ± standard deviation.

Results

Table 4.1 shows some characteristics of the study population, where Z_1 and Z_{100} are the measured impedances at 1 kHz and 100 kHz respectively. The usual differences between males and females were observed. ECW was highly correlated with H^2/R_e and with H^2/R_{tot} (0.89 and 0.87 ($p<0.0001$) respectively). After correction for TBW, ECW was still weakly correlated with H^2/R_e ($r = 0.08$, $p<0.05$). TBW was significantly correlated with H^2/R_{tot} and H^2/R_e (0.94 and 0.92 ($p<0.0001$) respectively), and remained so after correction for ECW (0.26 and 0.19 ($p<0.01$) respectively).

Prediction formulae for ECW and TBW were developed for each of the two subgroups, and applied to the other. Application of the group 1 formulae to group 2 and vice versa showed no differences between predicted and measured values. The data were therefore combined to provide general prediction formulae for ECW (Table 4.2) and TBW (Table 4.3).

Table 4.1. *Characteristics of the study sample*

	Males (n = 56)		Females (n = 47)	
	Mean	SD	Mean	SD
Age (years)	27.8	7.4	27.0	6.2
Weight (kg)	74.9	9.3	67.4*	11.8
Height (cm)	183.4	7.0	170.0*	5.9
Body mass index (kg/m^2)	22.2	2.2	23.4	4.7
Total body water (kg)	44.7	4.7	33.9*	3.7
Extracellular water (kg)	17.8	2.0	14.6*	1.6
R_e (Ω)	587	49	684*	73
R_{tot} (Ω)	386	36	464*	51
R_i (Ω)	1140	170	1462*	219
Z_1 (Ω)	587	53	667*	74
Z_{100} (Ω)	450	39	544*	61

*$p<0.001$, females compared with males.

Table 4.2. *Stepwise multiple regression of extracellular water (ECW) with H^2/R_e, weight, age and sex as independent variables for the total sample*

H^2/R_e (cm^2/Ω)	Weight (kg)	Age (years)	Sex male	Intercept (kg)	R^2	SEE (kg)
0.230	–	–	–	4.6	0.79	1.1
0.177	0.073	–	–	2.0	0.86	0.9

Age and sex did not enter into the equation ($p>0.05$).

Table 4.3. *Stepwise multiple regression of total-body water (TBW) with H^2/R_{tot}, weight, age and sex as independent variables for the total sample*

H^2/R_{tot} (cm^2/Ω)	Weight (kg)	Sex male	Age (years)	Intercept (kg)	R^2	SEE (kg)
0.428	–	–	–	7.0	0.88	2.4
0.368	0.133	–	–	2.1	0.90	2.1
0.289	0.163	2.4	–	4.7	0.91	2.0
0.266	0.195	2.8	– 0.09	6.4	0.92	2.0

The residuals of TBW were not significantly correlated with the mean value of predicted and measured TBW. The residuals of ECW were slightly but significantly correlated with the mean value of measured and predicted ECW (Fig. 4.3). The residuals of predicted ECW, but not the residuals of predicted TBW, were related to the ratio ECW/TBW (Fig. 4.4). The residuals of ECW were also correlated with the residuals of TBW (r = 0.57, $p<0.001$).

The formulae were applied to the data of five subjects before and 1–4 h

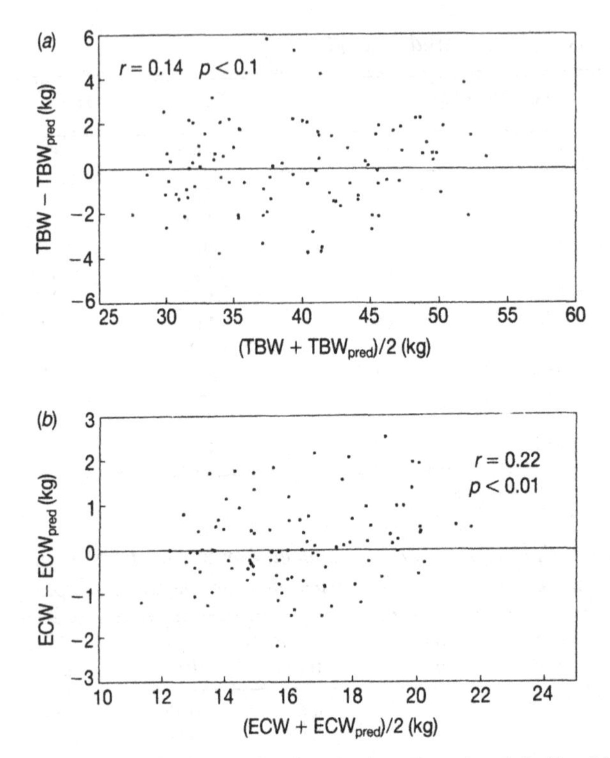

Fig. 4.3. Individual differences in measured and predicted (pred) total-body water (TBW) (*a*) and extracellular water (ECW) (*b*) in relation to the mean of predicted and measured water compartment.

after taking 40 mg frusemide. Predictions were based on just the impedance index H^2/R, to avoid weight loss being estimated as loss of TBW or ECW. Table 4.4 gives the losses of weight, urine and ECW calculated from urinary sodium, and the predicted losses of ECW and TBW, 3 h after taking the diuretic. At this time predicted ECW loss was lower than calculated ECW loss, whereas predicted TBW loss was higher than urine loss and weight loss. At 3 h the subjects drank 1 l of tap water, of which a mean of 0.2 l was voided 1 h later. The mean predicted losses in TBW and ECW at that time were 2.2 ± 0.5 kg and 1.1 ± 0.3 kg respectively, which differed significantly from the observed weight loss (0.8 ± 0.3 kg) and urine loss (1.6 ± 0.2 kg).

Discussion

In this study, body impedance has been calculated from a regression model relating impedance to frequency over the range 1 kHz to 100 kHz. Due to technical difficulties impedance at these extreme frequencies cannot be

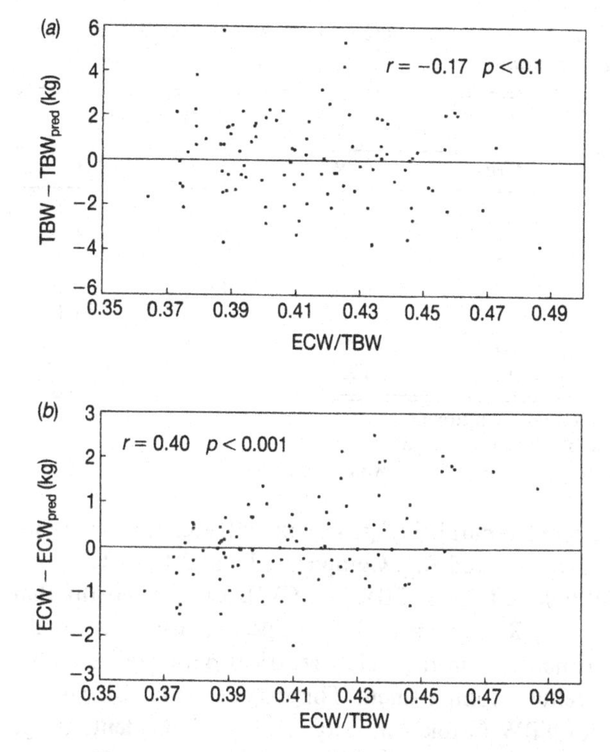

Fig. 4.4. Individual differences in measured and predicted (pred) total-body water (TBW) (*a*) and extracellular water (ECW) (*b*) in relation to the ratio of extracellular water to total body water.

measured reliably (Schwan, 1963; Settle *et al.*, 1980). The model fits the data very well, as shown by correlation coefficients exceeding 0.999 for all the subjects.

R_e and R_{tot} are theoretically strongly related to ECW and TBW respectively (Fig. 4.1). The correlation coefficient of the impedance index H^2/R with ECW and TBW was high, at both low and high frequency. This was due to the high correlation between ECW and TBW ($r = 0.93$, $p < 0.0001$). Therefore partial correlation coefficients were calculated, correcting for TBW or ECW in the correlation between impedance index and ECW or TBW respectively. The partial correlations showed that ECW was significantly correlated only with H^2/R_e, not H^2/R_{tot}. TBW remained significantly correlated with both H^2/R_e and H^2/R_{tot} after correction for ECW, but more highly with H^2/R_{tot}. The residual correlation of TBW with H^2/R_e after correction for ECW is understandable, as R_{tot} is partly determined by R_e (Fig. 4.1).

Assuming that the extracellular and intracellular compartments behave

Table 4.4. *Weight loss, urine loss, predicted ECW and predicted TBW loss (kg) 3 h after 40 mg frusemide*

	Loss of:			Predicted loss of:	
Case no.	Weight	Urine	ECW[a]	TBW	ECW
1	1.50	1.43	1.16	2.17	0.91
2	1.40	1.25	1.10	2.49	1.03
3	1.35	1.29	1.04	1.14	0.54
4	1.60	1.54	1.26	1.94	0.96
5	2.10	1.78	1.68	2.31	1.24
Mean	1.58	1.46	1.25	2.01*	0.96**
SD	0.30	0.21	0.25	0.53	0.25

[a] Calculated from sodium excretion in urine.
*$p = 0.07$ with urine loss, $p = 0.13$ with weight loss.
**$p < 0.001$ with weight and urine loss, $p < 0.05$ with ECW loss.

as two resistors connected in parallel (Fig. 4.1), the intracellular resistance R_i can be calculated from R_e and R_{tot}. Comparing R_e and R_i in Table 4.1 with ECW and ICW (where ICW = TBW − ECW), it is noteworthy that although ICW is larger, R_i is higher. So the specific resistivity of the intracellular compartment (or intracellular electrical pathway) is higher than that of the extracellular compartment. This may be the reason why the prediction formula for TBW (Table 4.3) is age and sex dependent, as age and sex partly determine the ratio of ECW to ICW.

The prediction formulae for ECW and TBW in Tables 4.2 and 4.3 have prediction errors of about 1 kg for ECW and 2 kg for TBW, which are comparable with those reported in the literature (Lukaski *et al.*, 1985; Kushner & Schoeller, 1986; Segal *et al.*, 1988, 1991; Deurenberg & Schouten, 1992). The predicted body water compartments did not differ (results not shown) from the measured compartments in males versus females or in subgroups based on body mass index (lower or higher than 25 kg/m^2) or age (younger or older than 30 years). The individual errors for both ECW and TBW were sometimes large, which may be unacceptable clinically. However, it should be recognised that the reference methods (deuterium oxide dilution and bromide dilution) also have errors. The relationship between the residuals of predicted ECW and the ratio ECW/TBW is less clear. It could be that although the current at low frequency does not pass through the cell membrane, there is some interaction between the membrane and the current, depending on the fibre direction (Rush *et al.*, 1963; Settle *et al.*, 1980), which lowers the measured resistance. At a low ECW/TBW ratio, hence relatively more ICW and more membrane, ECW could be overestimated. The correlation between the residuals of ECW and TBW may be due partly to errors in electrode

Table 4.5. *Changes in ECW and impedance values: theoretical considerations*

		Before	After	
			Calculated[a]	
R_e	(Ω)	587	⟶	684
R_i	(Ω)	1127		1127
R_{tot}	(Ω)	386		426
ECW[a]	(kg)	17.7	− −1.8 ⟶	15.9
TBW[b]	(kg)	44.3		40.8

[a] Using the formula $ECW = 0.22976 \times H^2/R_e + 4.6$
[b] Using the formula $TBW = 0.42836 \times H^2/R_{tot} + 7.0$

placement; for example, if the electrode is too proximal the impedance will be too low at all frequencies, overestimating both ECW and TBW. Also differences in body geometry (such as a different relative extremity length) can cause such differences.

When applied to the data of five subjects before and after taking 40 mg frusemide, the formulae failed to predict the loss of TBW accurately. Individually there were relatively large differences between weight loss and water loss on the one hand, and loss of ECW and TBW predicted from impedance on the other (Table 4.4). Marking the places where the electrodes were attached can reduce the variability in impedance measurements from about a 2–3% coefficient of variation to 0.5 % (unpublished data). At a group level impedance overestimated TBW loss, but slightly underestimated ECW loss. The results appear to contradict those of an earlier study where impedance was measured before and after a diuretic (Deurenberg & Schouten, 1992). However, re-analysing the data in the way described here – not at fixed frequencies, but with R_e and R_{tot} – made the results comparable.

The exaggerated loss in TBW can be explained theoretically. Consider a subject whose height, ECW, TBW, R_e and R_{tot} values are as given in Table 4.1 for males. R_i is calculated from R_e and R_{tot} as 1127 Ω. If a loss of 1.8 kg in ECW occurs, the theoretical changes in resistance (and hence in predicted water compartments) can be calculated (see Table 4.5). R_e changes from 587 to 684 Ω, but as no change in ICW occurs, R_i remains at 1127 Ω. Hence R_{tot} increases to 426 Ω. From this the loss of TBW after a loss of 1.8 kg ECW can be calculated to be 3.5 kg. The overestimate is due to the higher specific resistivity of the intracellular fluid compared with the extracellular fluid. When only (or mainly) ECW is lost, the specific resistivity of the total body fluid increases, and hence the increase in R_{tot} is greater than expected on the basis of water loss. Conversely when ICW is lost, the specific resistivity of the body fluid decreases, and the predicted change in TBW is underestimated.

This calculation shows that changes in body water which alter the ratio of ECW to ICW cannot be predicted from changes in impedance. The assumption that ECW and ICW behave as two resistors connected in parallel (Fig. 4.1) is probably too simplistic.

The slight underestimate in the loss of ECW (assuming that ECW excretion calculated from urinary sodium excretion is correct) could be due to a relatively lower loss of minerals compared with water after the use of frusemide. The amount of sodium excreted after 3 h was less than the normal homeostatic concentration (results not shown). Consequently the specific resistivity of the remaining extracellular fluid was lower, resulting in an underestimate of the ECW loss. However, the underestimate could also be due to the residuals of predicted ECW being related to the ratio ECW/TBW (Fig. 4.3).

In conclusion, multi-frequency impedance scans allow extracellular resistance and total resistance to be calculated with small estimation errors. Extracellular water can be predicted with an error of about 1 kg, using the impedance index H^2/R with extracellular resistance as the independent variable. TBW can be predicted with an error of about 2.2 kg using the impedance index at very high frequency. Body weight contributes significantly to the prediction of both ECW and TBW, and TBW also depends on age and sex. This may be a reflection of differences in the distribution of body water compartments between the sexes and between age groups. The individual prediction errors are probably too large for the impedance methodology to be useful in clinical practice, and the errors are likely to be even larger in less standardised measurement conditions.

Changes in body water induced by a diuretic drug were not predicted at either the individual or the group level. A valid prediction of changes in body water compartments is probably only possible when the change in body water reflects the initial ratio of ECW to ICW.

Acknowledgements

The statistical advice of Jan Burema, MSc, the technical assistance of Mr Frans J.M. Schouten, and the efforts of Catja Broekhoff, MSc, in discussing the results from the diuretic study are greatly appreciated.

References

de Lorenzo, A., Barra, P.F.A., Sasso, G.F., Battistini, N.C. & Deurenberg, P. (1991). Body impedance measurement during dialysis. *European Journal of*

Clinical Nutrition, **45**, 321–5.

Deurenberg, P. & Schouten, F.J.M. (1992). Loss of total body water and extra cellular water assessed by multi-frequency impedance. *European Journal of Clinical Nutrition*, **46**, 247–55.

Deurenberg, P., van der Kooy, K., Leenen, R. & Schouten, F.J.M. (1989). Body impedance is largely dependent on the intra- and extracellular water distribution. *European Journal of Clinical Nutrition*, **43**, 845–53.

Deurenberg, P., van der Kooy, K., Leenen, R., Weststrate, J.A. & Seidell, J.C. (1991). Sex- and age specific prediction formulas for estimating body composition from bio-electrical impedance: a cross validation study. *International Journal of Obesity*, **15**, 17–25.

Diaz, E.O., Villar, J., Immink, M. & Gonzales, T. (1989). Bioimpedance or anthropometry? *European Journal of Clinical Nutrition*, **43**, 129–37.

Forbes, G.B. (1987). *Human Body Composition*. New York: Springer-Verlag.

Forbes, G.B., Simons, W. & Armatruda, J.M. (1992). Is bio-electrical impedance a good predictor of body composition change? *American Journal of Clinical Nutrition*, **56**, 4–6.

Hoffer, E.C., Meador, C.K. & Simpson, D.C. (1969). Correlation of whole body impedance with total body water. *Journal of Applied Physiology*, **27**, 531–4.

Jenin, P., Lenoir, J., Roullet, C., Thomasset, A.L. & Ducrot, H. (1975). Determination of body fluid compartments by electrical impedance measurements. *Aviation, Space and Environmental Medicine*, **46**, 152–5.

Kooy, van der, K., Leenen, R., Deurenberg, P., Seidell, J.C., Westerterp, K.R. & Hautvast, J.G.A.J. (1992). Changes in fat-free mass in obese subjects after weight loss: a comparison of body composition measures. *International Journal of Obesity*, **16**, 675–83.

Kushner, R.F. & Schoeller, D.A. (1986). Estimation of total body water by bioelectrical impedance analysis. *American Journal of Clinical Nutrition*, **44**, 417–24.

Kushner, R.F., Kunigk, A., Alspaugh, M., Andronis, P.T., Leitch, C.A. & Schoeller, D.A. (1990). Validation of bioelectrical impedance analysis as a measurement of change in body composition in obesity. *American Journal of Clinical Nutrition*, **52**, 219–23.

Lukaski, H.C. & Johnson, P.P. (1985). A simple, inexpensive method of determining total body water using a tracer dose of D20 and infrared absorption of biological fluids. *American Journal of Clinical Nutrition*, **41**, 363–70.

Lukaski, H.C., Johnson, P.E., Bolonchuck, W.W. & Lykken, G.E. (1985). Assessment of fat-free mass using bio-electrical impedance measurements of the human body. *American Journal of Clinical Nutrition*, **41**, 810–17.

Miller, M.E. & Cappon, C.J. (1984). Anion exchange chromatographic determination of bromide in serum. *Clinical Chemistry*, **30**, 781–3.

Nijboer, J. (1970). *Electrical Impedance Plethysmography*, 2nd edn. Springfield, Illinois: C.C. Thomas.

Rush, S.R., Abildskov, J.A. & McFee, S. (1963). Resistivity of body tissues at low frequencies. *Circulation Research*, **12**, 40–50.

Schwan, H.P. (1963). Determination of biological impedances. In *Physical Techniques in Biological Research*, vol 6, ed. W.L. Nastuk. New York: Academic Press.

Segal, K.R., van Loan, M.D., Fitzgerald, P.I., Hodgdon, J.A. & van Itallie, T.B. (1988). Lean body mass estimation by bioelectrical impedance analysis: a four site cross validation study. *American Journal of Clinical Nutrition*, **47**, 7–14.

Segal, K.R., Burastero, S., Chun, A., Coronel P., Pierson R.N. & Wang, J. (1991). Estimation of extracellular water and total body water by multiple frequency bio-electrical impedance measurements. *American Journal of Clinical Nutrition*, **54**, 26–9.

Settle, R.G., Foster, K.R., Epstein, B.R. & Mullen, J.L. (1980). Nutritional assessment: whole body impedance and body fluid compartments. *Nutrition and Cancer*, **2**, 72–80.

SPSS/PC, V4.0 Manuals (1990). Chicago.: SPSS Inc.

Thomasset, A.L. (1962). Bio-electrical proporties of tissue impedance measurements. *Lyon Medical*, **201**, 101–18.

Vasquez, J.A. & Janoski, J.E. (1991). Validation of bioelectrical impedance analysis in measuring changes in lean body mass during weight reduction. *American Journal of Clinical Nutrition*, **54**, 970–5.

5 *Body composition assessed by electrical conductivity methods*

J.F. SUTCLIFFE AND M.A. SMITH

Introduction

Electrical conductivity offers the prospect of a rapid, safe, non-invasive, non-contact method for assessing body composition in humans. It was originally conceived in the USA in 1973 for the rapid assessment of the fat and lean proportions of meat packages, carcasses and livestock. The electronic meat measuring equipment (EMME) was adapted for human measurement in the early 1980s and renamed TOBEC. This operates on the principle that the impedance of a radiofrequency coil is changed when a human body is inserted and the change of impedance is related to the volume of body electrolytes. This was further developed in the mid 1980s to a scanning facility, in which the change of coil impedance as a function of the position of the subject was Fourier analysed to attempt to eliminate the influence of body shape on the estimate of fat-free mass. An alternative approach, named the tissue resonant impedance monitor (TRIM), was pioneered at the University of Swansea, Wales. The change of radiofrequency resulting from the insertion of the subject into a helical coil was related to the subject's lean body mass. This principle has been developed further at Leeds General Infirmary into a scanning facility driven with a free-running oscillator known as electromagnetic resonance (EMR). The average frequency shift that occurred when a subject was traversed completely through the coil was related to the subject's water space ($r = 0.87$). Subsequent analysis showed excellent correlation between frequency shift and surface area in calibration phantoms ($r = 0.998$), suggesting that only superficial induction of eddy currents was occurring and that the measurement was therefore relative to water at the surface.

Electronic meat measuring equipment (EMME)

In 1973, Wesley Harker patented an apparatus to measure the fat and lean proportions of meat, either as prepared packages, as carcasses, or in live animals. It was based on measuring the change of impedance of a coil, driven at a constant radiofrequency, as a result of passing the meat, carcass or live animal through the coil on a conveyor belt. The change of impedance of the coil is due to eddy currents induced in the conductive part of the tissue which tend to oppose the current in the coil. A sensing coil in a feedback circuit to the power amplifier was used to maintain the radiofrequency current in the main coil constant, so that the change of coil impedance manifested itself as a change of voltage amplitude and phase as the sample or animal passed through the coil. The coil was sufficiently long to enclose the sample or animal completely and the coil itself was completely enclosed by a radio frequency shield to minimise extraneous interference. The original coil was of rectangular cross-section (40 cm × 60 cm), sufficient to permit the passage of a 156 kg live animal, and comprised a series of spirally wound coils, each driven in phase from a coil drive distribution network. This arrangement was chosen to ensure that the electric field component was entirely in the transverse direction, while the magnetic component was entirely axial. The coil drive distribution system was also chosen because the coil could be several wavelengths long, for which a two-connection drive would be inappropriate. A range of frequencies (1–20 MHz) could be used to drive the coil system, though the chosen frequency was 5 MHz for a larger coil for live animals (SA-1) and 10 MHz for a smaller coil for the analysis of 27 kg boxes of beef (M-60).

Assuming the carcass/animal is a homogeneous conductive cylinder, of radius R, length L and volume V and mean electrical conductivity σ lying coaxially inside the coil, the change of impedance of the coil (δZ) is:

$$\begin{aligned}\delta Z &= \mathrm{K} \cdot \sigma \cdot (2\pi f)^2 \cdot R^4 \cdot L \\ &= K' \cdot \sigma \cdot (2\pi f)^2 \cdot (V)^2/L\end{aligned}$$

where f is the frequency and K and K' are constants. Hence:

$$V = K'' \cdot \sqrt{(\delta Z \cdot L/\sigma)}/(2\pi f)$$

where K' is a constant. This equation is only valid assuming that the electromagnetic field is not significantly disturbed by the presence of the carcass/animal in the coil, which is clearly untrue. Nevertheless this can give a qualitative indication of the variation of coil impedance with the size of the conductive volume of the subject. The strong dependence of the change of coil impedance on body size implies that the coil has to be calibrated for different species or meat package size. This geometric dependence has been verified by Khaled *et al.* (1985) using cylindrical phantoms. The equation suggests that higher frequencies will improve sensitivity, but this must be balanced by reduced penetration of electromagnetic waves at higher frequencies, such that the larger the body being measured, the lower the frequency employed.

The measurement of body potassium and conductivity in 75 live pigs and subsequent chemical analysis of the carcasses by Domermuth *et al.* (1976) in two experiments showed a reasonable correlation between the EMME number (impedance change) and body water ($r = 0.87$ and 0.68), protein ($r = 0.83$ and 0.60) and total body potassium (TBK) ($r = 0.75$ and 0.81). Similar results were also reported by Harrison & Van Italie (1982), who also stated that the radiofrequency energy flux was 0.01 mW/cm^2, one-thousandth of the upper exposure limit.

Total body electrical conductivity (TOBEC)

TOBEC 1

The original version of the apparatus used to measure total body electrical conductivity (TOBEC) in humans was essentially the same apparatus used for animals patented as EMME but was adapted for human dimensions and manufactured by a subsidiary of the Dickey-John Corporation of Auburn, Illinois, USA.

Presta and co-workers carried out the initial evaluation of TOBEC as a non-invasive method for human body composition assessment, determining total-body water (TBW) by dilution of ^3H_2O, TBK by counting ^{40}K and fat (from lean body mass, LBM) by skinfold anthropometry in 19 volunteers (4 male, 15 female). The following regression equations were derived (Presta *et al.*, 1983*a*):

$$\text{TBW (l)} = 30.214 + 0.954 \times \text{TOBEC No.} \quad (r = 0.87, \text{SEE} = 5.35\ \text{l})$$

$$\text{TBK (mmol)} = 1981 + 49.467 \times \text{TOBEC No.} \quad (r = 0.86, \text{SEE} = 291\ \text{mmol})$$

$$\text{LBM (kg)} = 38.479 + 1.081 \times \text{TOBEC No.} \quad (r = 0.693, \text{SEE} = 11.237\ \text{kg})$$

Reproducibility was better than 1% for four repeated TOBEC measurements of the same subject over 10 minutes. Subsequent measurement of LBM in 32 normal subjects (16 male, 16 female) by hydrodensitometry gave the relationship:

$$\text{LBM (kg)} = 18.323 + 5.1721 \times \sqrt{(\text{TOBEC No.} \times \text{HEIGHT}} + 3.816 \times \text{SEX}$$

where SEX = 0 for male, 1 for female ($r = 0.943$, SEE = 3.968 kg) (Presta *et al.*, 1983*b*). The TOBEC number also correlated well with body mass ($r = 0.942$). Similar measurements of TBW, TBK and LBM in 28 adolescents (21 male, 7 female) correlated best with $\sqrt{(\text{TOBEC No.})} \times \text{HEIGHT}$ ($r = 0.877$, $r = 0.859$ and $r = 0.911$ respectively) (Presta *et al.*, 1987).

TOBEC 2

In 1982 Van Loan & Mayclin introduced the TOBEC 2 apparatus. It comprised a cylindrical solenoidal coil, 79 cm in diameter and 185 cm long, driven from a 2.5 MHz constant current source. Whereas TOBEC 1 took a single measurement of coil impedance with the subject in the middle, TOBEC 2 gave 64 measurements of coil impedance (radio frequency voltage amplitude and phase) as the subject was traversed through the coil. The conductance of the coil (reciprocal of resistance) as a function of the position of the subject (FC0) was Fourier analysed (spatially) into two components (FC1 and FC2) in an attempt to eliminate the effect of body shape on the TOBEC measurement of conductive volume of the subject. Measurement of LBM by hydrodensitometry, TBW by dilution of ^2H_2O and TBK by counting ^{40}K in 40 normal subjects (20 male, 20 female) gave the following regression equations:

$$\text{LBM (kg)} = 22.999 + 0.1015 \times \text{FC0} + 0.0622 \times \text{FC1} - 0.2908 \times \text{FC2}$$
$$\text{TBW (l)} = 14.3 + 0.0544 \times \text{FC0} + 0.0701 \times \text{FC1} - 0.137 \times \text{FC2}$$
$$\text{TBK (mmol)} = 1288.54 + 0.8598 \times \text{FC0} + 18.078 \times \text{FC1} - 17.2168 \times \text{FC2}$$
$$\text{LBM(W) (kg)} = 20.694 + 0.0886 \times \text{FC0} + 0.0738 \times \text{FC1} - 0.2284 \times \text{FC2}$$

for which the respective correlation coefficients and standard errors of estimate are: $r = 0.991$, SEE = 1.429 kg; $r = 0.98$, SEE = 1.568 l; $r = 0.944$, SEE = 294.5 mmol; $r = 0.988$, SEE = 1.657 kg. LBM(W) is the lean body mass derived by hydrodensitometry, corrected for body water measured by isotopic dilution. This method of spatial Fourier analysis of the subject 'phase curve' was subsequently employed in all publications by this research group.

This pilot study was followed by a study of weight loss in 12 overweight women (Van Loan *et al.*, 1987*a*) and a cross-validation study of normal subjects from New York and California (103 and 54 respectively) (Van Loan *et al.*, 1987*b*). LBM was determined by hydrodensitometry in the latter study, from which the following regression equation in the combined populations was derived:

$$\begin{aligned}\text{LBM (kg)} = & -36.41 - (1.324 \times \text{SEX}) + ([0.01185 \times \sqrt{(\text{FC1})} \\ & \times \text{HEIGHT}] + [12.347 \times \sqrt{(\text{FC2})}] + (0.0627 \times \text{FC0}) \\ & - (0.9232 \times \text{FC2}) \quad (r = 0.98, \text{SEE} = 2.17\text{ kg})\end{aligned}$$

in which SEX = 0 for M, 1 for F. The mean difference in LBM between the two populations was 0.974 kg.

Cochran *et al.* (1988) reported measurements of TBW in 20 normal subjects (10 male, 10 female) by dual-isotope dilution (^2H and ^{18}O) and

TOBEC measurements. The resulting regression equation using ^{18}O to measure total body water was:

$$\text{TBW (l)} = 10.8 + (0.0724 \times \text{FC0}) - (0.221 \times \text{FC1}) + (0.0398 \times \text{AGE}) + (9.2 \times \text{HEIGHT} \times \text{AVERAGE LEAN CIRCUMFERENCE}) \quad (r = 0.997, \text{SEE} = 0.68 \text{ l})$$

where the average lean circumference is the average of measurements of body circumference at the chest, abdomen and thighs minus estimates of subcutaneous fat by skinfold anthropometry at these sites.

Subsequent papers by other investigators have employed the regression equations of Van Loan & Mayclin (1987) in their studies of other groups of normal subjects. This includes estimates of muscle mass by excretion of [^{3}H–methylhistidine and creatinine in 12 subjects by Horswill *et al.* (1989), a study of fat-free mass and TBW in 114 middle-aged and elderly subjects by Van Loan & Koehler (1990), measurements of fat-free mass, TBW and bone minerals in 50 teenagers by Van Loan (1990) and a study of body composition in 8 normal and 8 cystic fibrosis patients, matched for age, sex (6 male, 2 female) and height by Newby *et al.* (1990).

Paediatric TOBEC

Fiorotto and co-workers have developed TOBEC for the measurement of body composition in neonates and small children. The apparatus is an adaptation of the original equipment used to measure packages of meat but with a cylindrical coil. All versions operate at 10 MHz. The original apparatus had a sensitivity to meat packages > 1.5 kg and resolution of ± 35 g. The paediatric adaptations have improved resolution (± 22 g in an early model and ± 12 g in a newer version) and the latter has a lower sensitivity limit (> 1 kg) and a longer sensitive volume (100 cm versus 65 cm in HP-1). These developments have been reviewed by Fiorotto & Klish (1991).

Since the body of an infant conforms more closely to cylindrical geometry than the body of an adult, Fiorotto *et al.* (1987*a*) have attempted to validate formulas for the effect of body geometry and composition on the TOBEC measurements for which 12 miniature piglets and 35 adult rabbits were used as a model for human infants. These were chemically analysed for water, protein, fat mineral, sodium, chlorine and potassium content. On the theoretical basis that $\sqrt{(\text{TOBEC No.} \times \text{LENGTH})}$ is proportional to volume, the following regression equations were derived (for fat-free mass in the range 1.7–4.7 kg):

$$\text{TBW} = 0.0349 \times \sqrt{(\text{TOBEC No.} \times \text{LENGTH})} - 0.490 \quad (r = 0.998, \text{SEE} = 52\,\text{ml})$$

$$\text{FFM} = 0.0448 \times \sqrt{(\text{TOBEC No.} \times \text{LENGTH})} - 0.693 \quad (r = 0.998, \text{SEE} = 70\,\text{g})$$

and

$$\text{TBW} = -0.96 - 0.028 \times \sqrt{(\text{TOBEC No.} \times \text{LENGTH})} + 0.58 \times (\text{WEIGHT}/\text{LENGTH}^2) \quad (r = 0.999, \text{SEE} = 33\,\text{ml})$$

$$\text{FFM} = -1.35 + 0.036 \times \sqrt{(\text{TOBEC No.} \times \text{LENGTH})} + 0.82 \times (\text{WEIGHT}/\text{LENGTH}^2) \quad (r = 0.999, \text{SEE} = 42\,\text{g})$$

These regression equations were used to predict TBW and FFM in 34 healthy infants (Fiorotto *et al.*, 1987*b*), from which a regression equation relating FFM to anthropomorphic measurements was derived:

$$\text{FFM (kg)} = 0.12 \times \text{LENGTH (cm)} + 0.17 \times \text{WEIGHT (kg)} - 3.35$$

Cochran *et al.* (1989) extended the calibration work by expanding the extracellular fluid space of 11 miniature piglets causing an 8% increase in TBW. The piglets were chemically analysed after the TOBEC measurements, and regression equations for TBW and FFM are given.

Measurement of tissue resonance

An alternative approach in observing the interaction between the human body and an electromagnetic field is the measurement of the resulting change of frequency. This approach involves simpler instrumentation and may be inherently more precise than TOBEC because it is an entirely digital method.

Tissue resonant impedance monitor (TRIM)

The TRIM apparatus comprised a large helical coil, 1 m diameter and 1.8 m long, driven by a variable-frequency sinewave oscillator. As the frequency was increased, the coil exhibited anti-resonances in its impedance when the effective length of the windings became equal to an even number of quarter wavelengths. (This is illustrated for the EMR helical coil at Leeds in Fig. 5.2.) The sharpest anti-resonance was found to occur at the second harmonic, at which the effective length of the coil was equal to one wavelength. This occurs when the time for electricity to flow from one end of the coil winding to the other (at the speed of light) is equal to the period and phase of the alternating voltage applied across the coil. This occurred at 2.42 MHz.

When a human volunteer was moved to the midpoint of the coil, the frequency was manually readjusted back to the anti-resonance condition and the frequency change recorded. The change in impedance of the coil was also recorded prior to the readjustment of frequency. The correlation between the TRIM variable $\sqrt{(\%\text{ frequency shift} \times \text{height})}$ and LBM estimated by skinfold anthropometry (SKA), counting of ^{40}K (TBK), other anthropometric measurements (ANT) and bioelectrical impedance (BIA) in 45 normal subjects (22 male, 23 female) was $r = 0.92$, $r = 0.84$, $r = 0.96$ and $r = 0.87$ respectively. In a second group of 24 normal subjects (12 male, 12 female), additional measurements of TBW by dilution of ^3H_2O and total body nitrogen (TBN) by *in vivo* neutron activation analysis were carried out and estimates of LBM calculated. The respective correlation between (% frequency shift × height) and LBM estimated by SKA, TBK, ANT, BIA, TBW and TBN were $r = 0.95$, $r = 0.82$, $r = 0.99$, $r = 0.89$, $r = 0.92$ and $r = 0.86$ respectively (Chadwick & Saunders, 1990).

Electric field and magnetic field probes were used to measure the electromagnetic field within the coil at the first four harmonics (1.33 MHz, 2.42 MHz, 3.39 MHz and 4.30 MHz) and were found to agree well with theoretical predictions derived from Bessel functions (Chadwick, 1990).

Electromagnetic resonance (EMR)

The EMR facility at Leeds comprised a large helical coil (235 m of 8 mm minibore copper tubing wound as 50 turns onto a drum, 1.5 m in diameter and 2.4 m long) through which the supine subject could be completely traversed on a couch on wooden rails, 7 m long (Fig. 5.1). The variation of the impedance of the coil as a function of frequency is shown in Fig. 5.2.

The coil was driven by a free-running phase shift oscillator, the output of which was capacitative linked to the mid-point of the coil to lock into the second harmonic (approximately 1.57 MHz). This frequency corresponded to the length of the helix, being approximately 1.25 wavelengths. The resonant frequency was monitored by a frequency counter and sampled at 1 cm intervals of movement of the couch, yielding 500 measurements of frequency per complete traverse of the subject through the coil. These were summed in each direction and the total deducted from the sum of a corresponding set of frequency measurements taken during a traverse of the empty couch.

The frequency shift as a function of position of the subject showed a profile characterised by the intersection of the subject's body with the axial nodes of the standing wave at either end of the coil and the most sensitive node at its centre (Fig. 5.3). The asymmetry in the profile was due to different impedances at either end of the coil, one being linked to the

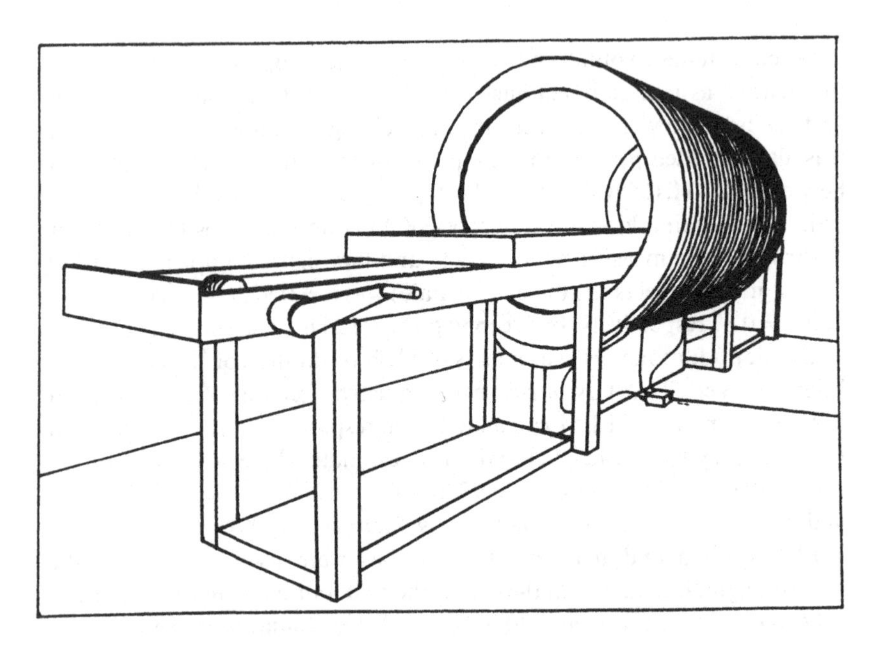

Fig. 5.1. The electromagnetic resonance facility (EMR) at Leeds General Infirmary.

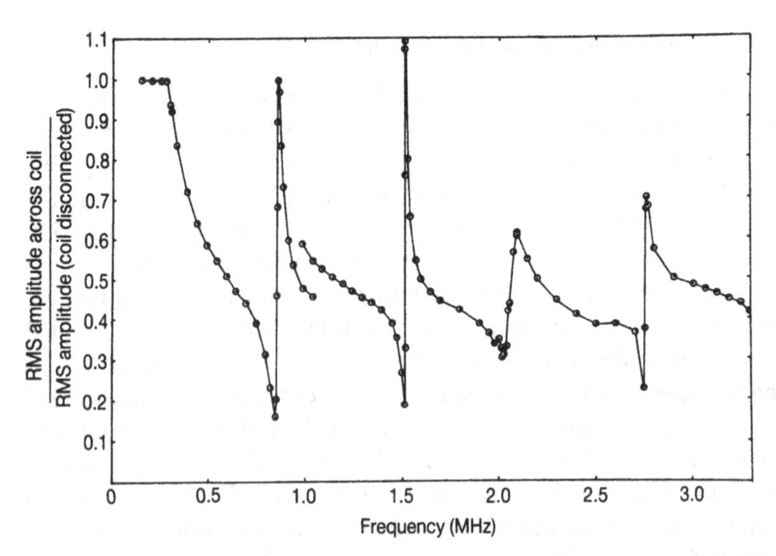

Fig. 5.2. The variation of the impedance of the helical coil with frequency. (The ratio of the voltage amplitude across the coil to the voltage amplitude with the coil disconnected as a function of frequency.) RMS, root mean squared.

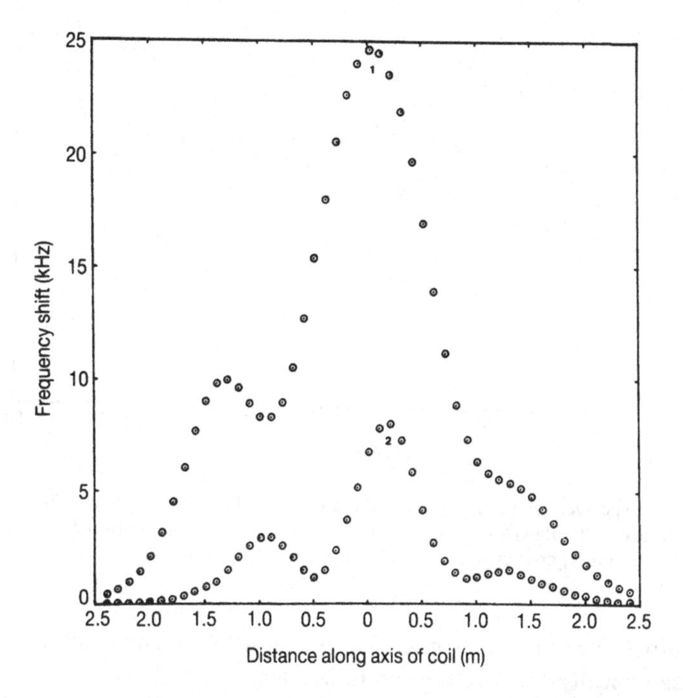

Fig. 5.3. The frequency shift as a function of position along the coil axis of a large (1) and a small (2) volunteer. Frequency was sampled at 10 cm increments. Volunteer 1 was a male of height 1.77 m, weight 76.5 kg, volunteer 2 was a female of height 0.96 m, weight 13.4 kg.

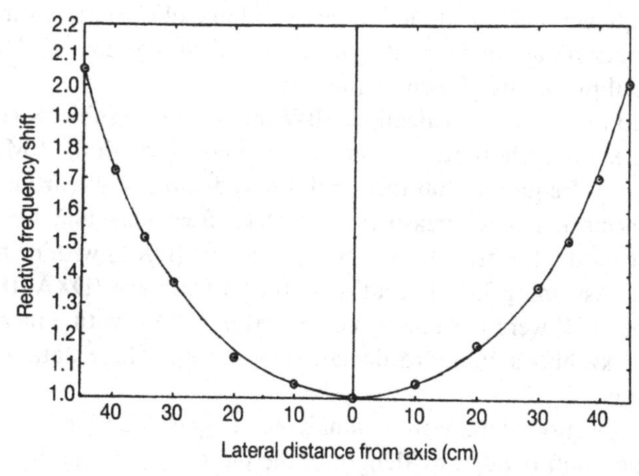

Fig. 5.4. The relative frequency shift as a function of lateral displacement from the coil axis of a single tube (length 1.75 m, volume 3 l) filled with 1% saline solution.

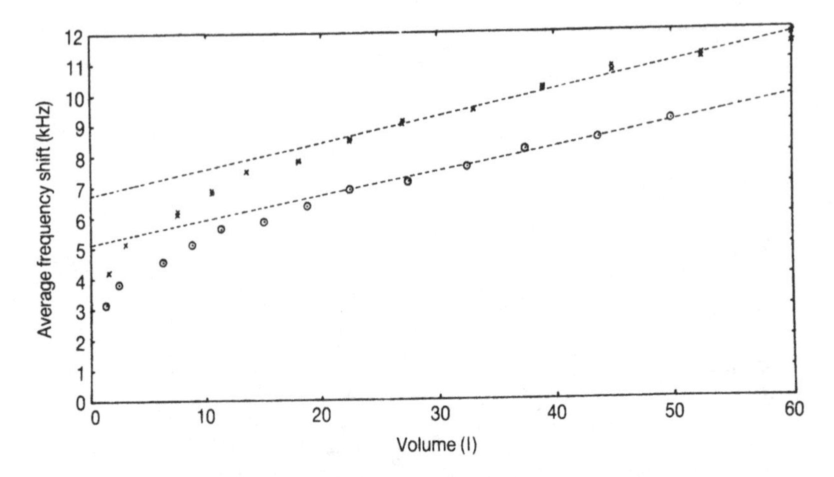

Fig. 5.5. The average frequency shift as a function of the volume of multiple-tube phantoms (1–40 tubes) filled with 1% saline solution. Circles, phantoms 1.5 m long (1.25 l per tube); crosses, phantoms 1.8 m long (1.5 l per tube).

oscillator output, the other to its input. The maximum frequency shift from an adult was typically 1.5% of the centre frequency.

The EMR helical coil was calibrated using tubular 'sausage' phantoms of two different lengths (1.8 m and 1.5 m), containing 1.5 l and 1.25 l of 1% saline solution respectively. The relative frequency shift as a function of the lateral displacement from the coil axis of a single tube is shown in Fig. 5.4. The average frequency shift as a function of the volume of a multiple-tube phantom (1–40 tubes) was approximately linear over the range 25–60 l. The results of the calibration are shown in Fig. 5.5.

This calibration was used to calculate TBW in 14 normal subjects (10 male, 4 female), each of whom was traversed five times through the EMR coil and the average frequency shift recorded. Reproducibility was around 1%. These subjects had had measurements of fat-free mass and bone mineral content by dual-energy X-ray absorptiometry (DXA) within the previous 2 years. Assuming 73% hydration of the fat-free mass (DXA) the two estimates of TBW were reasonably correlated ($r = 0.87$) with a mean difference of 2.2 kg and a standard deviation of 4.2 kg. These data are shown in Fig. 5.6.

When the calibration data were re-analysed, it was found that the average frequency shift divided by (length of the phantom)$^{1.5}$ was tightly correlated with the circumference of the phantom ($r = 0.998$). The circumference was calculated by two methods: either as the 'tapemeasure' distance, or as the external circumference of each of the outer layer of tubes. The data are shown in Fig. 5.7. This suggested that eddy currents were

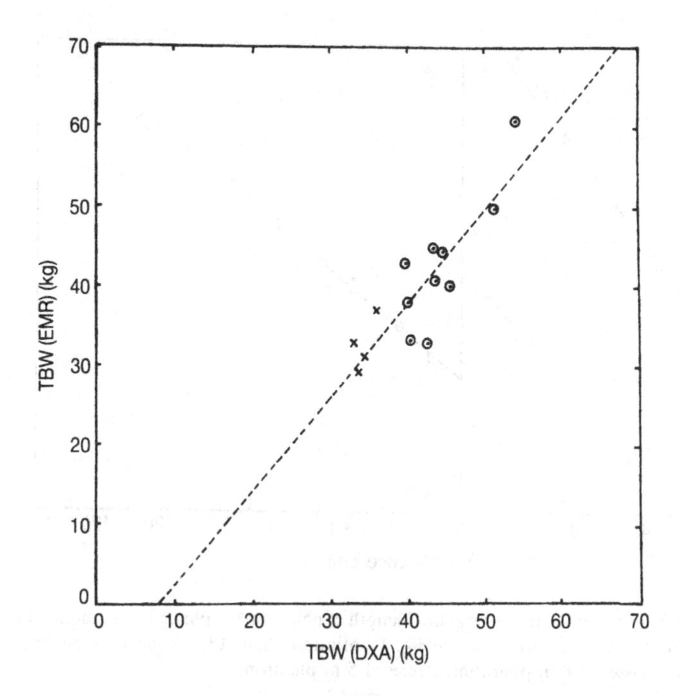

Fig. 5.6. The estimate of total-body water (TBW) from the EMR calibration as a function of TBW derived as 73% of fat-free mass measured by DXA in 14 normal subjects. Circles, males; crosses, females. (Mean (± SD) difference = 2.2 ± 4.2 kg, $r = 0.87$.)

induced only superficially in the phantom. When a multiple-tube phantom was supported in such a manner that the core could be withdrawn without disturbing the external geometry, no significant frequency shift was observed as a result of removing the core (6 tubes = 9 l). This corroborates the hypothesis of superficial induction of eddy currents.

The electromagnetic field inside a resonant helical coil differs from that in a TOBEC solenoid (a series of single or spiral loops driven in parallel) in having a predominant axial electric field component, whereas the transverse and circumferential electric fields predominate in the TOBEC solenoid. It is unclear whether this difference of configuration may cause the electromagnetic field to penetrate tissue less in an EMR/TRIM coil than in a TOBEC solenoid.

The *in vivo* results of measuring TBW using the EMR facility are very similar to the early TOBEC results (Presta *et al.*, 1983). Although the spatial Fourier analysis of the subject 'phase' curve was supposed to eliminate the effect of body geometry on the measurement of conductive volume by TOBEC 2 (Van Loan & Mayclin, 1987), Cochran *et al.* (1988) found it necessary to include an additional term in their regression equation involving lean body circumference and height. This is also suggested by the

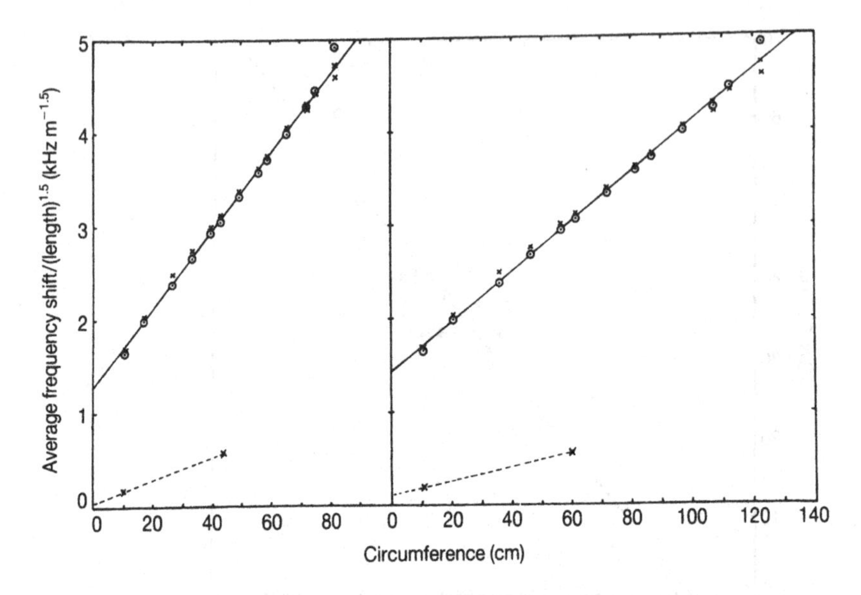

Fig. 5.7. The average frequency shift/(length of phantom)$^{1.5}$ plotted as a function of phantom circumference ($r = 0.9984$). Continuous line, 1% saline; dashed line, water; crosses, 1.8 m phantom; circles, 1.5 m phantom.

inclusion of a term (weight/height^2) to improve Fiorotto *et al.*'s calibration of the paediatric TOBEC (Fiorotto *et al.*, 1987*a*). Body mass, lean body mass, body water, surface area, height and circumference are all positively correlated, such that it is unclear what is being measured in a TOBEC facility.

Acknowledgements

The authors would like to acknowledge the assistance of Mr M. Whitaker with the construction of the EMR apparatus at Leeds and the Trustees of Leeds General Infirmary for financial support.

References

Chadwick, P.J. (1990). Studies of body composition by electromagnetic induction. PhD Thesis, University of Swansea.

Chadwick, P.J. & Saunders, N.H. (1990). TRIM, an electromagnetic body composition analyser. In *Advances in Body Composition Studies*, ed. S. Yasumura *et al.*, pp. 387–90. New York: Plenum Press.

Cochran, W.J., Wong, W.W., Fiorotto, M.L., Sheng, H.P., Klein, P.D. & Klish, W.J. (1988). Total body water estimated by measuring total body electrical conductivity. *American Journal of Clinical Nutrition*, **48**, 946–50.

Cochran, W.J., Fiorotto, M.L., Sheng, H.P. & Klish, W.J. (1989). Reliability of fat free mass estimates derived from total body electrical conductivity measurements as influenced by changes in extracellular fluid volume. *American Journal of Clinical Nutrition*, **49**, 29–32.

Domermuth, W., Veum, T.L., Alexander, M.A., Hedrick, H.B., Clark, J. & Eklund, D. (1976). Prediction of lean body composition of live market weight swine by indirect methods. *Journal of Animal Science*, **43**, 966–76.

Fiorotto, M.L. & Klish, W.J. (1991). Total body electrical conductivity measurements in the neonate. *Clinics in Perinatology*, **18**, 611–27.

Fiorotto, M.L., Cochran, W.J., Funk, R.C., Sheng, H.P. & Klish, W.J. (1987*a*). Total body electrical conductivity measurements: effects of body composition and geometry. *American Journal of Physiology*, **252**, R794–800.

Fiorotto, M.L., Cochran, W.J. & Klish, W.J. (1987*b*). Fat free mass and total body water of infants estimated from total body electrical conductivity measurements. *Pediatric Research*, **22**, 417–21.

Harker, W. (1973). US Patent No. 3735247, Method and apparatus for measuring fat content in animl tissue either *in vivo* or in slaughtered and prepared form. EMME Co. assignee.

Harrison, G.G. & Van Italie, T.B. (1982). Estimation of body composition: a new approach based on electromagnetic priciples. *American Journal of Clinical Nutrition*, **35**, 1176–9.

Horswill, C.R., Geeseman, R., Boileau, R.A., Williams, B.T., Layman, D.K. & Massey, B.H. (1989). Total body electrical conductivity (TOBEC): relationship to estimates of muscle mass, fat-free weight and lean body mass. *American Journal of Clinical Nutrition*, **49**, 593–8.

Khaled, M.A., McCutcheon, M.J., Canlas, J. & Butterworth, C.E. (1985). Effects of body geometry on TOBEC measurement. *Proceedings of the Southern Biomedical Conference*, **4**, 171–5.

Newby, M.J., Keim, N.L. & Brown, D.L. (1990). Body composition of adult cystic fibrosis pateints and control subjects as determined by densitometry, bioelectrical impedance, total body electrical conductivity, skinfold measurements and deuterium oxide dilution. *American Journal of Clinical Nutrition*, **52**, 209–13.

Presta, E., Wang, J., Harrison, G.G., Bjorntorp, P., Harker, W. & Van Italie, T.B. (1983*a*). Measurement of total body electrical conductivity: a new method for estimation of body composition. *American Journal of Clinical Nutrition*, **37**, 735–9.

Presta, E., Segal, K.R., Gutin, B., Harrison, G.G. & Van Italie, T.B. (1983*b*). Comparison in man of total body electrical conductivity and lean body mass derived from body density: validation of a new body composition model. *Metabolism*, **32**, 524–7.

Presta, E., Casullo, A.M., Costa, R., Slonim, A. & Van Italie, T.B. (1987). Body composition in adolescents: estimation by total body electrical conductivity. *Journal of Applied Physiology*, **63**, 937–41.

Van Loan, M.D. (1990). Assessment of fat-free mass in teenagers: use of TOBEC methodolgy. *American Journal of Clinical Nutrition*, **52**, 586–90.

Van Loan, M.D. & Koehler, L.S. (1990). Use of total body electrical conductivity for the assessment of body composition in middle aged and elderly individuals. *American Journal of Clinical Nutrition*, **51**, 548–52.

Van Loan, M.D. & Mayclin, P. (1987). A new TOBEC instrument and procedure for the assessment of body composition: use of Fourier components to predict lean body mass and total body water. *American Journal of Clinical Nutrition*, **45**, 131–7.
Van Loan, M.D., Belko, A.Z., Mayclin, P. & Barbieri, T.F. (1987*a*). Use of total body electrical conductivity for monitoring body composition changes during weight reduction. *American Journal of Clinical Nutrition*, **46**, 5–8.
Van Loan, M.D., Segal, K.R., Bracco, E.F., Mayclin, P. & Van Italie, T.B. (1987*b*). TOBEC methodology for body composition assessment: a cross validation study. *American Journal of Clinical Nutrition*, **46**, 9–12.

6 *Body composition in malnutrition*

P.S. SHETTY

Introduction

A reduction in body weight is an invariable and constant feature of inadequate food intake or any restriction of access to food both in animals and in humans. This decrease in body weight is associated with changes in the composition of the residual tissues of the body. The changes in body composition do not bear a simple relation to the changes in body weight, since other factors such as changes in hydration occur along with differential mobilisation of tissues, and these determine the nature of the body compositional changes in humans during food restriction. This chapter therefore examines the changes in body water and body fluid compartments as well as changes in body fat and the non-fat components. Differential alterations in the loss of active tissue or body cell mass as related to contributions from muscle and non-muscle tissue are examined along with changes in organ mass where data are available. This review aims to examine changes that occur in the following four conditions associated with malnutrition in humans: (1) experimental human semi-starvation, (2) malnutrition associated with disasters such as famines or wars, (3) chronic undernutrition in adults and (4) malnutrition in children.

Body composition changes during experimental semi-starvation in adults

Decrease in body weight is the most obvious manifestation of inadequate food intake in humans. However, the relationship between the degree and duration of inadequate energy intake and body weight changes is not a simple one, since important body compositional changes occur over a period of time which affect the proportion of fat stores as well as altering the level of hydration of the body. The loss of body weight during energy restriction involves loss of variable proportions of fat and other tissue material including protein and minerals, along with changes in the extracellular fluid compartment, with the body at the same time tending to be relatively overhydrated (Grande, 1964). The best-documented changes in body composition that occur as a result of experimental semi-starvation in human adults are those reported by Keys *et al.* (1950) from the

Table 6.1. *Changes in body composition following experimental semi-starvation in 32 male adults*

	Control		Experimental semi-starvation			
			Week 12		Week 24	
	kg	% of body wt	kg	% of body wt	kg	% of body wt
Body weight	69.5	100	56.9	100	53.6	100
Fat mass[a]	9.7	13.9	5.0	8.7	2.8	5.2
Active tissue mass[b]	39.9	57.5	32.0	56.2	29.2	55.5
Bone mineral[c]	2.8	4.0	2.8	4.9	2.8	5.3
Thiocyanate space[d]	17.1	23.5	17.2	30.3	17.9	34.0

Data compiled from the Minnesota semi-starvation study of Keys *et al.* (1950).
[a] Fat mass was computed from densitometric measurements using underwater weighing.
[b] Active tissue mass was calculated as the total body weight less the sum of fat mass, bone mineral and thiocyanate space.
[c] Bone mineral was assumed to be 4% of the normal or control body weight
[d] Thiocyanate space is an estimate of extracellular fluid space and is expressed as kilograms of protein-free plasma.

Minnesota semi-starvation experiment. An earlier experimental semi-starvation study by Benedict *et al.* (1919) in two groups of adults provided no data on body composition changes. In the Minnesota study, the average loss of body weight among a group of 32 male subjects who were semi-starved amounted to 24.2% of their initial body weight at the end of 24 weeks of semi-starvation, a loss of 16.8 kg (Table 6.1).

Body fluid compartments

During the Minnesota experiment (Keys *et al.*, 1950), plasma volumes of the subjects were estimated at regular intervals using the dye-dilution technique; the relative mass of the blood cells was estimated by the haematocrit method. The blood volume reduced in absolute terms from 5.83 l to 5.38 l around week 23 of semi-starvation – a reduction of about 7.7%. The plasma volume increased (+8.9%) while the cell volume reduced, resulting in a considerable degree of anaemia. The relative volumes of plasma and blood expressed per kilogram body weight rose over the period of semi-starvation. After 22 weeks of semi-starvation, body weights were reduced by 23.0%, the plasma volume per kilogram body weight was 40.9% greater than its initial value, while the total blood volume increased by only 19.6% per kilogram body weight over the same period.

Estimates of extracellular body fluid using the thiocyanate method showed that the thiocyanate space during weeks 22–24 of semi-starvation averaged 34.0% of body weight as compared with values of 23.6% obtained in normal, young controls. Semi-starvation seemed to be associated with a large and dramatic increase in the relative hydration of the body. The average excess hydration was estimated at 6.25 l of thiocyanate space at 24 weeks, the excess extracellular fluid accounting for 11.7% of the body weight at that time. In the Minnesota study it was assumed that the thiocyanate space minus the plasma volume was a measure of the extravascular space including the interstitial fluids, and hence a general index of tissue hydration. Since both the extracellular fluid volume and the plasma volume expressed per kilogram body weight increased during semi-starvation, it was concluded that the hydration of both intravascular and extravascular compartments had increased.

Bone density

Densitometric measurements of bone during the Minnesota semi-starvation experiment also led to the conclusion that a loss of 24% of body weight over a period of 24 weeks of semi-starvation was not accompanied by generalised demineralisation of the bones. It was assumed that about 4% of the normal initial body weight was made up of bone mineral and that this mass remained unaltered throughout the period of semi-starvation (Table 6.1).

Adipose tissue, active tissue mass and organ size

The total amount of adipose tissue estimated on the basis of specific gravity measurements in the Minnesota study showed that fat expressed as a percentage body weight reduced from an initial value of 13.9% to 5.2% by week 24 of semi-starvation. This amounted to an average individual adipose tissue loss of 6.9 kg, i.e. a reduction of about 70% from initial levels.

The 'active tissue mass', defined as the difference between the body weight and the sum of fat mass, thiocyanate space and bone mineral mass, reduced by 27 per cent over the period of 24 weeks (Table 6.1). The changes in the proportion of the active tissue mass to the total body weight were not very great compared with the changes in the proportions of the body fat content. Estimates of alterations in organ size were also attempted during the Minnesota study. Measurements of cardiac size and volume using roentgengrams showed a 12% reduction in the transverse cardiac diameter and an 18% reduction in cardiac volume after 24 weeks of semi-starvation.

Body composition changes in human adults during food restriction associated with natural disasters, famine and war

The changes in body weight and body composition that accompany starvation and food restriction in adult humans have been extensively documented. Many of the data have been compiled and reviewed in the classic volumes on the *Biology of Human Starvation* by Keys and others (1950). The effects of complete starvation have been documented in professional fasters (Morgulis, 1923). Severe degrees of food restriction that occurred during disasters such as famines and wars, as well as that suffered by prisoners of war and inmates of concentration camps, have provided most of the information on body weight and body composition changes in humans. Considerable amounts of information related to the latter have been obtained from autopsies carried out following deaths associated with such disasters.

Body fluid compartments

A universal finding during severe underfeeding associated with food restriction during famine and war is a relative increase in the water content of the body. The disproportionate changes in body water associated with the reduction in body weight and loss of tissue from the body result in a relative increase in hydration. Eventually, this increase in body hydration is clinically recognisable as puffiness of the face and ankles and may ultimately progress to generalised oedema. This oedema of starvation, which is referred to as 'famine oedema', 'hunger oedema' or 'war oedema', is not a universal phenomenon seen in all individuals who have limited access to food.

Estimates of blood and plasma volume made on inmates of concentration camps liberated at the end of World War II (Mollison, 1946) indicated a reduction in blood volume which was less than the proportion expected on the basis of the estimated weight loss. The relative blood volume per kilogram body weight appeared to be elevated and so was the plasma volume; this was associated with a considerable degree of anaemia. Beattie and others (1948) provided extensive data on plasma volumes and extracellular fluid volumes in two groups of severely undernourished Dutch and German adults. There was little, if any, change in the absolute plasma volume. But the relative plasma volume per unit of body mass was increased in proportion to the degree of body weight loss. The same was true of the extracellular volume measured as thiocyanate space; the absolute thiocyanate space was only slightly expanded but represented 34.2% of body weight compared with a mean value of 23.5% reported in normal men (Henschel *et al.*, 1947). Walters *et al.* (1947) likewise concluded that there was little change in the absolute values for total blood volume of ex-prisoners of war in a state of severe undernutrition, despite the marked body weight loss seen.

Body weight and body fat content

The reduction in body weight seen in individuals during food restriction has been estimated variously depending on the duration of restriction. Weight loss was a maximum of 15% during a 5 month siege of troops in World War I (Hehir, 1922), between 22% and 26% among Russians and Ukrainians during the Russian famine of 1920–2 (Ivanovsky, 1923), 28.4% during a famine in India (Martin, 1928) and between 9.3% and 13.6% among Parisian civilians during the years of World War II (Trémolières, 1948). There are several other reports documenting various degrees of initial body weight loss during food restriction. During complete starvation for 40 days in professional fasters, weight loss was 25.3% of initial body weight (Morgulis, 1923). However, the maximum recorded was 41% of body weight loss by a male who refused all food and died on the 63rd day of his total fast (Meyers, 1917).

Although estimates of body fat content are not available, autopsy studies indicate a total loss of subcutaneous adipose tissue along with depletion in other fat depots reported by Schittenhelm & Schlecht (1918) during autopsies on 200 undernourished patients. Several other studies have commented on the total loss of subcutaneous adipose tissue with depletion of pericardial and perirenal depots as well as omental fat (Park, 1918; Simonart, 1948).

Muscle tissue and organ mass

Extreme muscle wasting and atrophy is a prominent feature of severe undernutrition and starvation in humans. Autopsies carried out on 359 starvation victims of the Warsaw ghetto showed severe muscle atrophy in over 61% of subjects, while 36.3% showed a moderate degree of atrophy (Stein, 1947). Starvation produces loss of weight in all organs in the body except the central nervous system (Jackson, 1925). With the exception of the brain, all the organs examined, including the heart, liver, spleen and kidney, were on average considerably reduced in weight as compared with those of normal individuals (Table 6.2). Summaries of autopsy data on famine victims (Porter, 1889) and inmates of concentration and prison camps (Uehlinger, 1948) also showed marked reduction in the weight of

Table 6.2. *Proportions (expressed as % of normal) of various organ masses at autopsy in severely malnourished children and adults*

	Brain	Heart	Liver	Spleen	Kidneys
Malnourished children[a]	90	60	–	–	80
Malnourished adult victims of Warsaw ghetto[b]	107	80	54	52	75
Malnourished adult victims of concentration camps[c]	90	70/75	70	–	90

[a] Kerpel-Fronius & Frank (1949).
[b] Steub (1946).
[c] Uehling (1948).

internal organs such as heart, liver, kidneys, spleen and pancreas, with minimal change in weight of the brain.

Body composition changes in chronic undernutrition in adults

There is relatively little information about body composition changes in chronically undernourished adults. Although there have been some earlier reports (Gopalan *et al.*, 1953; Holmes *et al.*, 1956) the best documented data have been obtained from male subjects aged 18–56 years in Colombia, South America (Barac-Nieto *et al.*, 1978).

Total-body water and body fluid compartments

Table 6.3 summarises the data on total-body water (TBW) and body fluid compartments, i.e. plasma volumes, intra- and extracellular compartments, in undernourished adults of varying grades of nutritional deficiency (Barac-Nieto *et al.*, 1978). The results for the mildly undernourished group are comparable to values obtained from normal adults for almost all parameters studied (Edelman *et al.*, 1952). A careful examination of the data of Barac-Nieto and colleagues (1978) shows that the adults in this group had mean body weights of 52 kg and mean body mass indices (weight/height^2) of 21.4, within the range of normality. Hence the data can be assumed to reflect normal values for body composition in this group of individuals. In the severely undernourished TBW was reduced, although when expressed as a percentage body weight it was significantly increased. The extracellular fluid volumes were also not altered in absolute terms in the moderately and severely undernourished adults. However, the extracellular volume expressed as a percentage of body weight or TBW was significantly increased as undernutrition progressed. Gopalan *et al.* (1953)

Table 6.3. *Body water compartments in adults with progressive chronic undernutrition*

	Grade of adult undernutrition		
	Normal or mild	Moderate	Severe
Total body water[a]			
litres	31.4	28.3	26.7
% body wt	60.3	58.5	62.4
Extracellular water[b]			
litres	14.1	13.9	14.4
% body wt	27.3	28.6	33.8
% TBW	45.4	49.2	54.3
Plasma volume[c]			
litres	2.6	2.7	2.5
% body wt	5.0	5.7	5.8
% TBW	8.2	9.7	9.3
Intracellular water[d]			
litres	17.2	14.5	12.1
% body wt	33.0	29.9	28.6
% TBW	54.6	50.8	45.6

Compiled from data of Barac–Nieto *et al.* (1978)
[a] Total-body water (TBW) was estimated by dilution using tritiated water.
[b] Extracellular fluid (ECF) volume was estimated from the thiocyanate distribution space.
[c] Plasma volume was estimated by the dilution of Evans blue in plasma.
[d] Intracellular fluid volume was calculated as TBW – ECF.

reported an excess accumulation of fluid in this compartment in adults with severe undernutrition and nutritional oedema. They measured plasma volume, extracellular fluid volume using thiocyanate, and TBW using urea in 14 severely undernourished adults. The absolute plasma volumes (2.07 l) were lower, although as a proportion of the rest of the tissues the values were higher. The TBW at 39.7 l accounted for 83.3% of body weight while the mean extracellular fluid volume was 22.1 l (46.3% of body weight) and the mean intracellular fluid volume was 17.6 l (37.0% of body weight). Both the TBW and extracellular fluid volumes were higher in absolute terms and relative to the body weight, and remained considerably higher even after the oedema was clinically cleared. Holmes *et al.* (1956) also reported high values for TBW and for thiocyanate space in 8 malnourished African patients, which subsequently approached normality following nutritional rehabilitation. It thus appears that there is a marked increase in the proportion of extracellular water within the TBW pool in chronic undernutrition even in the absence of oedema (Widdowson, 1985).

Plasma volume was not altered in absolute terms, but when expressed relative to TBW or body weight it increased progressively as undernutrition advanced (Table 6.3) (Barac-Nieto *et al.*, 1978). On the contrary, intracellular water was reduced both in absolute terms and when expressed as a percentage of TBW or body weight. In the severely undernourished group, the intracellular water decreased out of proportion to the decrease in body weight or TBW, representing a disproportionately larger deficit of intracellular water (Table 6.3). These significant reductions in intracellular water content probably represent deficits in cell mass in undernutrition; the deficits in cell mass are further underestimated if one takes into account the increased hydration of cells which is known to occur in undernutrition (Holmes *et al.*, 1956).

Body fat and fat-free mass

Chronically undernourished adults have reduced stature, probably the result of growth deficits associated with childhood malnutrition. Despite their shorter stature along with their low body weights they also have low body mass indices (weight/height^2) (Shetty, 1984), well below the accepted cut-off of 18.5 (James *et al.*, 1988). The body fat content is reduced and reported values vary from about 10–12% (Soares & Shetty, 1991) to as low as 6% (Shetty, 1984). Estimates of fat-free mass also indicate a marked reduction in the proportion of active cell mass in these adults. Barac-Nieto *et al.* (1978) have computed the body cell mass and cell solids in undernourished adults and reported that both are significantly reduced in moderate and severe undernutrition, reflecting a deficit in body cell mass (Table 6.4). The same group concluded that undernutrition was characterised by a loss in body fat, a loss in body cell mass and possibly a small deficit in extracellular solids.

Comparisons of the contribution of muscle cell mass and non-muscle or visceral cell mass to the fat-free mass have also been made in chronically undernourished adults (Table 6.4). The data of Barac-Nieto *et al.* (1978) have shown that with increasing severity of undernutrition, the reduction in the fat-free mass is contributed to by an early loss of muscle cell mass. The reduction in muscle cell mass was in proportion to the decrease in total-body cell mass with a moderate degree of undernutrition, and as undernutrition progressed muscle cell mass decreased out of proportion to the loss of body cell mass while visceral mass was only minimally affected. Body cell mass reduced by 29% while muscle cell mass decreased by 41% in severe undernutrition and visceral mass changed by only 1.4% (Table 6.4). More recent estimates of body composition in chronically energy-deficient adults also support the fact that there is a greater reduction of muscle mass, with visceral mass apparently being spared (Soares *et al.*, 1991). These changes are important since they may explain the adaptive reduction in basal metabolic rate seen in undernourished adults (Shetty, 1993).

Table 6.4. *Changes in body composition of adults with progressive, chronic undernutrition*

	Grade of adult undernutrition		
	Normal or mild	Moderate	Severe
Body weight[a]			
kg	52.0	48.2	42.5
Fat-free mass[a]			
kg	42.8	38.3	36.1
Body cell mass[b]			
kg	24.1	20.3	17.1
% body wt	46.1	41.9	40.3
Muscle mass[c]			
kg	17.2	14.6	10.1
% body wt	33.1	30.3	23.8
Visceral mass[d]			
kg	6.9	5.7	7.0
% body wt	13.0	13.4	14.5

Data compiled from Barac-Nieto *et al.* (1978).
[a] Fat-free mass was estimated from total body water measurements.
[b] Body cell mass was derived from estimates of intracellular water and their relationship to body cell mass.
[c] Muscle mass was estimated from 24 h urinary creatinine excretion assuming 1 g of creatinine corresponds to 20 kg muscle inclusive of 16% extracellular fluid, i.e. 1 kg of muscle free of ECF = 60 mg of 24 h creatinine excretion.
[d] Visceral mass was calculated as body cell mass minus muscle mass.

Body composition changes during malnutrition in children

A severely malnourished child is not only a smaller sized individual for his or her age, but also manifests marked body composition changes related to proportions of different tissues as well as their chemical composition. Changes in organ size occur and their relative contribution to non-fat or lean tissue is altered. The changes not only reflect the degree of severity of the protein-energy malnutrition (PEM) the child suffers from, but also vary with the type of PEM, i.e. kwashiorkor, marasmus or marasmic kwashiorkor.

Body water and body fluid compartments

TBW measurements have been made in infants and children suffering from severe PEM as well as during and following recovery from PEM as a result of nutritional rehabilitation (Smith, 1960; Alleyne, 1967). Isotopic dilution has been the most favoured technique for *in vivo* measurements in children. Measurements of body water have also been made directly during whole-body analysis of infants and children *post mortem*. The results of isotopic dilution and direct body compositional analysis tended to suggest that in both kwashiorkor and marasmus the body was overhydrated and that TBW was increased (Garrow *et al.*, 1968). Hansen *et al.* (1965) reported an inverse relationship between TBW (expressed as a percentage body weight) and the body weight deficit of the malnourished infant (expressed as a percentage of expected weight for age): the greater the weight deficit, the higher the TBW content of the body. There seemed to be no difference according to whether the child had kwashiorkor or marasmus. Marked changes in TBW were shown to occur during the clinical response of the malnourished child to nutritional rehabilitation. Loss of oedema and recovery from PEM restored the TBW proportion of body weight to that in normal children. Only during the early phases of nutritional repletion did the children show a marked tendency to increase the TBW content of the body.

It has generally been assumed that much of the increase in TBW in childhood PEM is largely extracellular as in the case of adults who were semi-starved experimentally or suffered the consequences of severe food restriction. Estimates of the extracellular fluid compartment based on measurement of thiocyanate space have confirmed this: extracellular water was increased in absolute terms and when expressed as a percentage of body weight (Alleyne *et al.*, 1977). An inverse relationship between extracellular water and body weight has also been demonstrated – a relationship not different from that between TBW and body weight (Alleyne, 1967). Intracellular water content of tissue samples (muscle, liver, skin) obtained at biopsy in both marasmic children and those with kwashiorkor (with or without oedema) showed an increase in water content (Frenk *et al.*, 1957; Smith, 1960; Vis *et al.*, 1965). However, values based on muscle biopsies are highly variable and therefore do not confirm that intracellular water content is increased in relation to tissue solids in childhood PEM (Waterlow, 1992).

Changes in muscle and organ mass

Due to the marked changes in body water content and its distribution, as well as alterations in the concentration of electrolytes such as potassium within cellular tissue, estimates of body fat mass and fat-free or lean body mass based on the classical methods such as TBW or total body potassium are unlikely to provide reliable data. Hence estimates of lean body mass and fat mass are not available. However, measurement of the structural and cellular components of lean tissue have been made. Measurement of the collagen and non-collagen nitrogen of the lean tissues has shown that the collagen contributes a much larger proportion of the total body protein in children with PEM (i.e. 36–48%) than in normal children (27%) (Picou *et al.*, 1966). It appears that in severe PEM the collagen protein content is maintained despite considerable reduction in the total lean tissue mass.

Muscle mass is severely depleted in childhood malnutrition and the catabolism of muscle tissue is greatly increased in PEM. Marasmic infants who have body weights 50% of normal for their age have a total muscle mass which is only 30% of normal. Comparisons of sections of whole sartorius muscle showed that a 12-month-old child dying of PEM had a similar total transverse section of sartorius muscle to that of a fetus at 31 weeks of gestation, and an area a tenth that a of a normal child of 12 months (Montgomery, 1962). The sizes of muscle bundles and individual fibres were reduced proportionately. Individual muscle fibres were so attenuated in severe PEM that the number of fibres in a given area in a 12-month-old marasmic infant was no different from that in a 36-week fetus. Cheek and his colleagues (1970) showed that the reduction in muscle mass was due mainly to a reduction in cell size and cell mass, and not to differences in cell numbers.

The weight and size of the heart are reduced in severe PEM, and so is the size of the kidneys in children dying of acute PEM. Wasting of organs such as the pancreas and the gastrointestinal tract is also well recognised. The weight and size of the liver depend on the presence and extent of fatty infiltration, resulting in what is described as 'fatty liver disease' (Waterlow, 1948) or 'fatty liver hepatomegaly' (Alleyne *et al.*, 1977). The brain and the central nervous system are perhaps the only organ systems to be spared. Kerpel-Fronius & Frank (1949) reported that the weight of the brain was relatively well preserved in children dying of severe malnutrition despite dramatic reductions in the weight of most internal organs of the body (Table 6.2).

Conclusions

The body compositional changes that accompany moderate to severe malnutrition in adults and children show similar characteristics. The marked reduction in body weight is associated with both loss of body fat stores and depletion of lean tissues that constitute the metabolically more active tissue mass. In the lean tissue compartment there is a tendency to disproportionate loss of skeletal muscle tissue as opposed to organ size and mass, although both are reduced. There are marked changes in the body water compartments with a universal tendency to overhydration of the body. The extracellular fluid compartment relative to body weight expands disproportionately even in the absence of oedema. The features of body compositional change that occur in adult victims of food restriction during famine, wars and natural disasters are similar to those in adults who are experimentally semi-starved.

References

Alleyne, G.A.O. (1967). The effect of severe protein calorie malnutrition on the renal function of Jamaican children. *Pediatrics*, **39**, 400–11.

Alleyne, G.A.O., Hay, R.W., Picou, D.I., Stanfield, J.P. & Whitehead, R.G. (1977). *Protein-Energy Malnutrition*. London: Edward Arnold.

Barac-Nieto, M., Spurr, G.B., Lotero, H. & Maksud, M.G. (1978). Body composition in chronic undernutrition. *American Journal of Clinical Nutrition*, **31**, 23–40.

Beattie, J., Herbert, P.H. & Bell, D.J. (1948). Famine oedema. *British Journal of Nutrition*, **2**, 47–65.

Benedict, F.G., Miles, W.R., Roth, P. & Smith, H.M. (1919). In *Human Vitality and Efficiency under Prolonged Restricted Diet*. Washington: Carnegie Institute, publ. no. 280.

Cheek, D.B., Hill, D.E., Cordano, A. & Graham, G.G. (1970). Malnutrition in infancy: changes in muscle and adipose tissue before and after rehabilitation. *Pediatric Research*, **4**, 135–44.

Edelman, I.S., Olney, J.M., James, A.H., Brooks, L. & Moore, F.D. (1952). Body composition studies in human beings by dilution principles. *Science*, **115**, 447–9.

Frenk, S., Metcoff, J., Gómez, F., Ramos-Galvan, R., Cravioto, J. & Antonowicz, I. (1957). Composition of tissues. *Pediatrics*, **20**, 105–20.

Garrow, J.S., Smith, R. & Ward, E.E. (1968). *Electrolyte Metabolism in Severe Infantile Malnutrition*. Oxford: Pergamon.

Gopalan, C., Venkatachalam, P.S. & Srikantia, S.G. (1953). Body composition in nutritional edema. *Metabolism*, **2**, 335–43.

Grande, F. (1964). Man under caloric deficiency. In *Handbook of Physiology, Adaptation to the Environment*. pp. 911–37. Washington, D.C.: American Physiological Society.

Hansen, J.D.L., Brinkman, G.L. & Bowie, M.D. (1965). Body composition in protein-calorie malnutrition. *South African Journal of Nutrition*, **1**, 33–7.

Hehir, P. (1922). Effects of chronic starvation in the siege of Kut. *British Medical Journal*, i, 865–8.

Henschel, A., Mickelsen, O., Taylor, H.L. & Keys, A. (1947). Plasma volume and thiocyanate space in famine edema and recovery. *American Journal of Physiology*, **150**, 170–80.

Holmes, E.G., Jones, E.R., Lyle, M.D. & Strainer, M.W. (1956). Malnutrition in African adults: effect of diet on body composition. *British Journal of Nutrition*, **10**, 198–219.

Ivanovsky, A. (1923). Physical modifications of the population of Russia under famine. *American Journal of Physical Anthropology*, **6**, 331–53.

Jackson, C.M. (1925). *The Effect Of Inanition and Malnutrition upon Growth and Structure*. Philadelphia: Blakiston.

James, W.P.T., Ferro-Luzzi, A. & Waterlow, J.C. (1988). Definition of chronic energy deficiency in adults. Report of Working Party of IDECG. *European Journal of Clinical Nutrition*, **42**, 969–81.

Kerpel-Fronius, E. & Frank, K. (1949). Einige besonderheiten der korperzusammensetsung und wasserverteilung bei der sauglingsatrophie. *Annales Paediatrica*, **173**, 321–30.

Keys, A., Brozek, J., Henschel, A., Mickelson, O. & Taylor, J.L. (1950). *The Biology of Human Starvation*. Minneapolis: University of Minnesota Press.

Martin, R. (1928). *Lehrbuck der Anthropologie: I Somatologie*. Jena: Fischer.

Meyers, A.W. (1917). Some morphological effects of prolonged inanition. *Journal of Medical Research*, **36**, 51–78.

Mollison, P.L. (1946). Observations on cases of starvation at Belsen. *British Medical Journal*, i, 4–8.

Montgomery, R.D. (1962). Muscle morphology in infantile protein malnutrition. *Journal of Clinical Pathology*, **15**, 511–21.

Morgulis, S. (1923). *Fasting and Undernutrition*. New York: Dutton.

Park, F.S. (1918). War edema. *Journal of the American Medical Association*, **70**, 1826–7.

Picou, D., Halliday, D. & Garrow, J.S. (1966). Total body protein, collagen and non-collagen protein in infantile protein malnutrition. *Clinical Science*, **30**, 345–51.

Porter, A. (1889). *The Diseases of the Madras Famine of* 1877–78. Madras: Government Press.

Schittenhelm, A. & Schlecht, J. (1918). Uber odemkrankheit mit hypotonischer bradykardie. *Berliner Klinische Wochenschrift*, **55**, 1138–42.

Shetty, P.S. (1984). Adaptive changes in basal metabolic rate and lean body mass in chronic undernutrition. *Human Nutrition: Clinical Nutrition*, **38**, 573–81.

Shetty, P.S. (1993). Chronic undernutrition and metabolic adaptation. *Proceedings of the Nutrition Society*, **52**, 267–84.

Simonart, E.F. (1948). *La Denutrition de Guerre. Etude Clinique, Anatomopathologique et Therapeutique*. Brussels: Acta Medica Belgica.

Smith, R. (1960). Total body water in malnourished infants. *Clinical Science*, **19**, 275–85.

Soares, M.J. & Shetty, P.S. (1991). Basal metabolic rates and metabolic efficiency in chronic undernutrition. *European Journal of Clinical Nutrition*, **45**, 363–73.

Soares, M.J., Piers, L.S., Shetty, P.S., Robinson, S., Jackson, A.A. & Waterlow, J.C. (1991). Basal metabolic rate, body composition and whole body protein turnover in Indian men with differing nutritional status. *Clinical Science*, **81**, 419–25.

Stein, J. (1947). Pathological anatomy of hunger disease. In: *Hunger disease* (translated from Polish and reprinted), ed. M. Winick, pp. 207–34. New York: Wiley.

Trémolières, J. (1948). Les enquêtes nutrition de 1940 à 1947. *Bulletin: Institut National de l'Hygiene* (Paris), **3**, 225–50.

Uehlinger, E. (1948). Pathologische anatomie der hungerkrankheit und des hungerödems. *Helvetica Medica Acta*, **14**, 584–601.

Vis, H., Dubois, R., Vanderborght, H. & De Maeyer, E. (1965). Etude des troubles électrolytiques accompagnant le kwashiorkor marastique. *Revue Francais des Etudes Cliniques Biologie* **10**, 729–41.

Walters, J.H., Rossiter, R.J. & Lehmann, H. (1947). Electrolyte disturbances accompanying marasmic kwashiorkor. Malnutrition in Indian prisoners of

war in the Far East. *The Lancet*, i, 205–10.
Waterlow, J.C. (1992). Body composition and body water. In *Protein-Energy Malnutrition*, ed. J.C. Waterlow, pp. 26–39. London: Edward Arnold.
Widdowson, E.M. (1985). Responses to deficits of dietary energy. In *Nutritional Adaptation in Man*, ed. K. L. Baxter & J.C. Waterlow, pp. 97–104. London: John Libbey.

7 *Influence of body composition on protein and energy requirements: some new insights*

C.J.K. HENRY

Introduction

A major development in recent years has been the introduction of new techniques and methods for the estimation of body composition (Kral & van Itallie, 1993). This is further illustrated by the varying and innovative methods that have been described in the previous chapters. In concert with these advances has been an interest in understanding the relationships between body composition on the one hand, and its impact on human biology and function on the other. The purpose of this chapter is to describe relationships between body composition and human nutrient requirements. In other words, we wish to explore the question: 'Does body composition influence our protein/energy requirements?' It is useful and indeed appropriate in human biology to examine protein and energy metabolism together for the following reasons: (1) protein metabolism, notably nitrogen balance, is profoundly influenced by energy balance, so that any perturbance in energy balance significantly alters protein metabolism (Calloway & Spector, 1954); (2) it is now conventional to express protein requirements as a proportion of total energy requirements, namely protein: energy percentage (P:E %) (Payne, 1975).

Historical background

It may be argued that Sarrus & Rameaux in 1839 were the first to suggest a link between energy metabolism and body composition, i.e. body size, when they proposed a relationship between heat loss and surface area. Following their lead, several classical physiologists such as Rubner (1883) and Richet supported this view and the principle was passed down as the 'surface law of heat loss'. Further analysis and advances in measuring surface area in various animals led to the 'surface law' being questioned (Kleiber, 1932). More significantly, the 'surface area rule' could not provide a convincing biological or physiological explanation for heat *production* in homeotherms.

It was Benedict & Talbot (1921) who proposed the concept of active protoplasmic mass and related it to basal metabolic rate. Their proposal was the first attempt to relate body composition *per se* to energy metabolism, a concept which had, for the first time, some *physiological* and *biological meaning.* It was Behenke's work (1953) that introduced the concept of lean body mass (LBM), and laid the foundation of the first two-compartment model of body composition. He later showed that LBM was closely correlated to energy metabolism. Since that time, several investigators have repeatedly shown a strong relationship between LBM and energy metabolism in a range of animals (von Dobeln, 1956; Elia 1992). This relationship is now a well-established concept in biology.

Whilst there has been extensive and long-term interest in the relationship between body composition and energy metabolism, little research has been conducted on the likely role body composition may play in protein metabolism, and hence in determining protein requirements. This chapter, therefore, will concentrate on exploring in more detail the role body composition may play in protein metabolism.

Analysis

We can begin our analysis by first examining whether protein and energy metabolism show any quantitative relationship with each other. We can then explore the impact of body composition on protein metabolism. An appropriate starting point is to investigate the relationship between protein and energy metabolism in an extreme situation of total food restriction, namely total starvation. It is an occasion when external food sources do not interfere with our analysis. During starvation, the body stores of protein and fat are the only fuels available and they are therefore mobilised for both the protein and energy needs of the body. Although it is almost impossible to conduct long-term studies on starvation today, remarkably, several studies of prolonged total starvation in animals and man were conducted at the turn of this century. They may be used to detail any quantitative relationships between protein and energy metabolism. Fig. 7.1 shows the results of the classic study by Benedict (1915) on his subject Levanzin who fasted for 30 days. Nitrogen loss in the urine – fasting urinary nitrogen loss (FUNL) – may be taken as a proxy for protein metabolism (breakdown), and the basal metabolic rate (BMR) as a proxy for total energy expenditure. A reduction or fall in BMR is one of the most obvious and consistent observations during total starvation (Fig. 7.1). What has been less well appreciated is the parallel fall in FUNL during the same period of starvation. FUNL, which represents protein metabolism (expressed as g/day), and BMR, which represents energy breakdown (expressed as

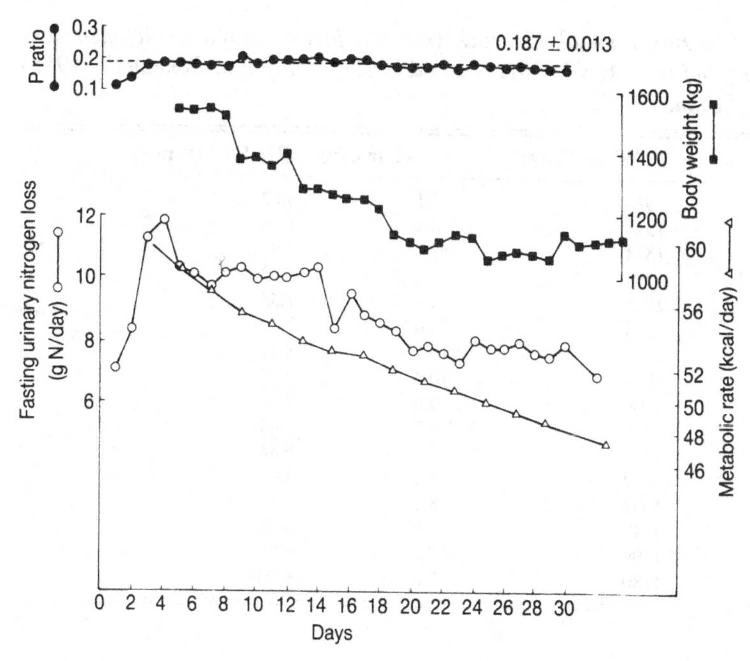

Fig. 7.1. Fasting urinary nitrogen loss, metabolic rate and P ratio during a 31 day fast in a normal weight male subject.

kcal/day), may be used to determine whether protein and energy metabolism show any close quantitative relationship. This may be done by dividing FUNL by BMR. Such an analysis is shown in Table 7.1. The results indicate a remarkable *constancy* of FUNL/BMR during the entire period of starvation, showing little variation between the early and later stages. The constancy of FUNL/BMR at about 7 mg N/kcal is in marked contrast to the fall in BMR and nitrogen loss of about 30% during the entire period of starvation.

Although Benedict's study is unique it is not the only study available and we have collected from the literature other starvation studies conducted by Japanese and Dutch investigators. In all human subjects FUNL/BMR was remarkably constant over varying periods of starvation (Table 7.2).

Whilst all the human studies surveyed show a constancy of FUNL/BMR during starvation, it is of biological interest to explore whether such a phenomenon is also apparent in other animals. A review of starvation studies conducted in the rat and guinea pig (Table 7.3) shows that these animals also exhibit a close quantitative relationship between protein and energy metabolism. These results, taken together, demonstrate for the first

Table 7.1. *Basal metabolic rate (BMR), fasting urinary nitrogen loss (FUNL) and the FUNL/BMR ratio during starvation in Benedict's* (1915) *subject Levanzin*

Day of fast	BMR (kcal/day)	FUNL (g/day)	FUNL/BMR ratio
1	1536	7.1	4.62
2	1543	8.4	5.44
4	1514	11.9	7.86
6	1394	10.2	7.32
8	1406	10.3	7.33
10	1284	10.0	7.79
12	1246	10.1	8.11
14	1214	10.4	8.57
16	1106	9.6	8.68
18	1116	8.3	7.44
20	1126	7.7	6.84
22	1097	7.7	7.02
24	1087	8.1	7.45
26	1147	7.9	6.89
28	1109	7.6	6.85
30	1130	7.8	6.90

Table 7.2. *Nitrogen loss and metabolic rate in normal-weight subjects during starvation*

Subject no.[a]	Days of fasting	Initial body weight (kg)	Fasting urinary nitrogen loss, FUNL (g/day)[b]	Basal metabolic rate, BMR (kcal/day)[b]	FUNL/BMR
1	31	60	7.8	1130	6.9
2	30	58	7.9	951	8.3
3	26	79	7.5	1176	6.3
4	17	51	7.9	950	8.3
5	16	43	10.3	951	10.8
6	12	50	10.7	1087	9.8
7	10	57	9.5	1508	6.3
Mean ± SD					8.1 ± 1.7

[a] 1, Benedict (1915); 2–6, Takahira (1925); 7, van Noorden (1907).
[b] Measured on same day of starvation.

time that a quantitative relationship exists between FUNL and BMR in a range of homeotherms.

Closer examination of Table 7.2 reveals two important features: (1) there is a constancy of FUNL/BMR in an individual during starvation; (2) there is a wide variation in FUNL/BMR *between* individuals during starvation. This is a point we shall return to later in the chapter.

Table 7.3. *Nitrogen loss and metabolic rate during total fasting in rats and guinea pigs*

Day of fast	Body weight (g)	Fasting urinary nitrogen loss, FUNL (mg/day)	Basal metabolic rate, BMR (kcal/day)	FUNL/BMR	P ratio
Rat					
2	115	85	16	5.4	0.13
6	100	55	11.3	4.91	0.12
7	96	62	11.3	5.4	0.13
8	93	62	11.4	5.4	0.13
9	87	70 (82%)	11.4	6.1	0.15
Mean ± SD				5.4(±0.43)	0.13
Guinea pig					
2	625	417	102	4.0	0.10
3	582	395	89	4.4	0.11
4	550	332	77	4.3	0.11
5	524	332	73	4.5	0.11
6	498	343	75	4.5	0.11
9	428	294 (70%)	69 (67%)	4.2	0.10
Mean ± SD				4.3(±0.19)	0.106(±0.005)

Data adapted from Benedict & Fox (1934) for rat and Morgulis (1923) for guinea pig.

Constancy of tissue mobilisation

A possible theory to accommodate the observed relationship between FUNL and BMR can be derived from the model developed by Payne & Dugdale (1977), initially proposed as a model for energy regulation. These authors suggested that the way in which individuals regulate energy balance depends on their P ratio, defined as:

$$\text{P ratio} = \frac{\text{Energy stored or mobilised as protein}}{\text{Total energy stored or mobilised}} \qquad (7.1)$$

Subjects with a high P ratio will deposit their excess food energy as protein which has a high maintenance energy requirement, and thereby maintain energy balance more effectively than those with a low P ratio who will tend to deposit excess food energy as fat.

A direct relationship between protein and energy is also suggested by the P ratio theory. During tissue mobilisation, it predicts that a constant fraction of energy should be derived from the breakdown of protein, i.e. the P ratio should remain constant. If the assumption is made that body protein is 16% nitrogen and has a metabolisable energy value of 4 kcal (16.7 kJ)/g (Henry, 1984), then the P ratio during fasting may be calculated, using the fasting FUNL and fasting metabolic rate, or BMR, as follows:

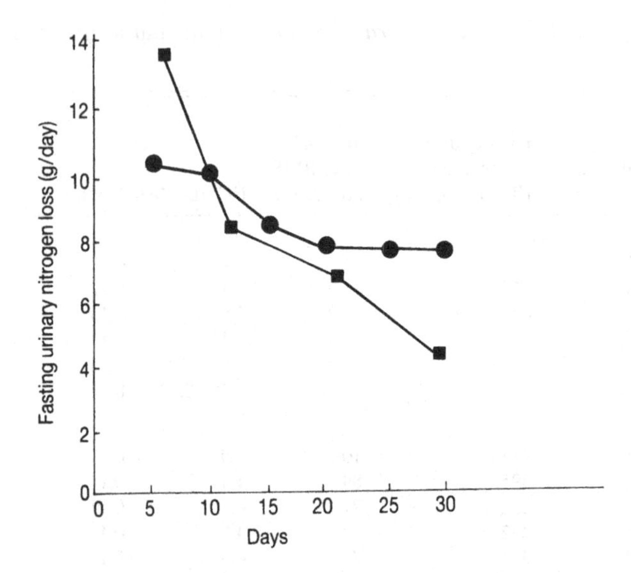

Fig. 7.2. Mean fasting urinary nitrogen loss in normal-weight (circles) and obese (square) subjects.

$$\text{P ratio} = \frac{\text{Fasting urinary nitrogen loss} \times 6.25 \times 16.7}{\text{Fasting metabolic rate BMR}} \qquad (7.2)$$

As is evident, the P ratio is a *derived* function of the ratio FUNL/BMR discussed earlier. By merely converting FUNL to energy units, the numerator and denominator are mathematically normalised (see equation 7.2).

Tissue mobilisation in the obese

A closer examination of the pattern of tissue mobilisation during starvation in 'obese' and 'normal' subjects reveals some interesting features. Fig. 7.2 shows the pattern of urinary nitrogen loss during a 31 day fast in a normal-weight subject (60.6 kg) studied by Benedict (1915) and compares it with that of an obese subject (154 kg) studied by Gilder *et al.* (1967) and starved for a comparable period. After the first week, the obese subject lost less fasting urinary nitrogen than the normal-weight subject.

This difference in nitrogen loss is further highlighted if we compare the FUNL/BMR ratio of normal subjects (Table 7.2) with that of obese subjects (Table 7.4). The obese have a lower FUNL/BMR ratio and appear to mobilise about 5% of their energy from protein breakdown, in contrast

Table 7.4. *Nitrogen loss and metabolic rate in obese subjects during starvation*

Subject no.	Days of fasting	Initial body weight (kg)	Fasting urinary nitrogen loss, FUNL (g/day)	Basal metabolic rate, BMR (kcal/day	FUNL/BMR
1	38	123	3.9	1985	1.9
2	38	132	7.9	1707	2.3
3	38	132	7.5	1773	1.8
4	21	154	7.9	1740	2.5
5	21	190	4.56	2480	1.8
6	14	166	3.41	2100	1.6
7	21	124	5.7	2328	2.4
Mean $\pm$ SD					2.04 ± 0.35

Table 7.5. *Relationship between body composition and FUNL/BMR ratio in different species*

Animal species	FUNL/BMR
Mouse	5.7
Rat	6.4
Rabbit	6.5
Sheep	6.5
Human, normal weight	5.8
Guinea pig	4.5
Pig	3.8
Human, obese	2.5
Rat, obese (VMH)	1.7
Rat, obese (cafeteria)	2.3
Rat, obese (Zucker)	0.9

From Henry (1984).

to normal weight subjects who mobilise approximately 20% of their energy from protein breakdown. (This may be calculated using equation 7.2), i.e. the P ratio). It therefore appears that normal-weight and obese subjects respond differently during starvation, and that the adipose tissue exerts a profound influence on protein mobilisation. The FUNL/BMR ratio in normal-weight subjects (Table 7.2) ranged from 6.3 to 10.8 mg N/kcal – markedly different values. It therefore appears that the tendency to mobilise protein as a fuel source during starvation is much lower in the obese than in normal-weight subjects. Put another way, it may be said that the obese lose a smaller amount of nitrogen per basal kilocalorie (smaller FUNL/BMR) than normal-weight subjects during starvation and the adipose tissue exerts a significant influence on protein mobilisation. Are these observations unique to man? It appears not. Table 7.5 shows a collation of FUNL/BMR values during starvation in a range of animals

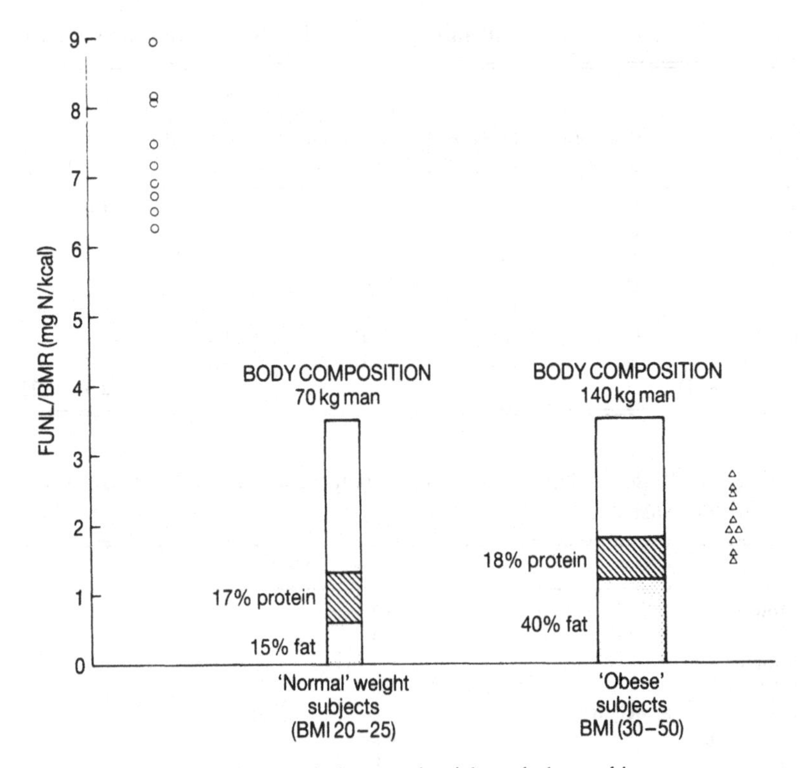

Fig. 7.3. FUNL/BMR ratio in normal-weight and obese subjects.

with varying degrees of adiposity. Two points emerge from this table: (1) irrespective of the species, the greater the adiposity the *lower* the FUNL/BMR ratio; (2) human obesity and animal obesity appear to show similar patterns of tissue mobilisation during starvation. Fig. 7.3 shows this in a schematic form for human subjects.

In summary, body composition, notably adipose tissue, appears to have a strong influence on the FUNL/BMR ratio. The *greater* the amount of adipose tissue, the *smaller* the FUNL/BMR ratio in animals and man.

Allometric analysis

The studies discussed thus far have been confined to humans and a few animal species. Are the observations recorded in these animals applicable to other homeotherms? This may be explored using the technique of allometry.

Allometry was a technique introduced by Huxley (1932) for relating the size of components of the animal's body to total body size. The generalised equation may be written as:

$$M = aW^b \tag{7.3}$$

where M is the size of organs, a is a constant, b the exponent and W the weight. Equation (7.3) may be written as follows:

$$\log M = \log a + b \log W \tag{7.4}$$

Following Huxley, Brody (1945) and Kleiber (1932) independently showed an allometric relationship for homeotherms between body weight and basal metabolic rate of the form:

$$\text{BMR (kcal/day)} = 70W^{0.75} \tag{7.5}$$

Henry (1984) analysed the allometry of FUNL in 12 mammalian species ranging from the mouse to the cow with a weight range of 0.022–453 kg, with the following results:

$$\text{FUNL (mg N/day)} = 418W^{0.77} \tag{7.6}$$

Since the weight exponent in both cases is equal to or close to 0.75, we can derive the following relationship between FUNL and BMR, by dividing equation (7.6) by (7.5):

$$\text{FUNL} = 6 \text{ mg N/basal kcal} \tag{7.7}$$

This interspecific generalisation suggests that in most normal-weight homeotherms during starvation, the *average* loss of nitrogen is approximately 6 mg N/kcal. More significantly the mean FUNL/BMR value derived from allometry reported here is not very different from the values for FUNL/BMR observed in normal-weight human subjects (Table 7.2), confirming the overall validity of equations (7.5) and (7.6).

The major elements described in the previous sections may be summarised as follows:

1. The FUNL/BMR ratio in an individual is constant during starvation.
2. A quantitative relationship between protein metabolism and energy metabolism has been established in homeotherms. During tissue mobilisation (i.e. starvation) this is of the form FUNL/BMR = 6 mgN/kcal.
3. Whilst FUNL/BMR is constant in an individual, it varies widely between individuals.
4. Obese subjects have a considerably smaller FUNL/BMR ratio than normal-weight subjects.
5. Body composition, notably adipose tissue, appears strongly to dictate protein mobilisation (i.e. FUNL) during starvation.

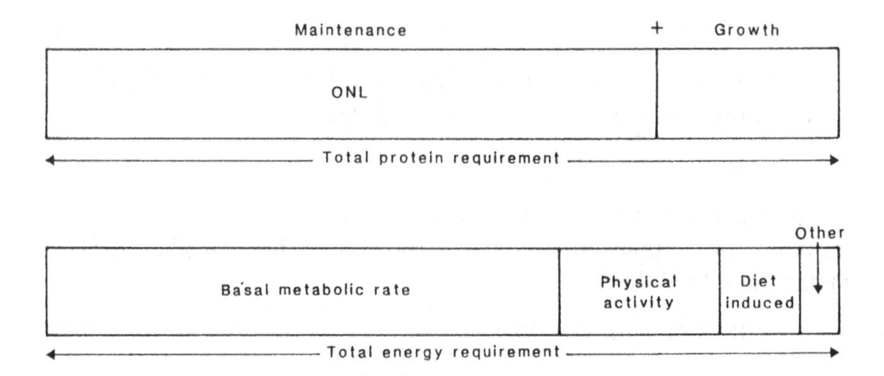

Fig. 7.4. Factorial breakdown of protein and energy requirements into their components.

Body composition and human protein requirements

How can the observations discussed thus far be used to explore the relationship between body composition and protein requirements? Of the several methods available for estimating protein requirements, the factorial method has been widely used to determine protein requirements in children and adults (FAO, 1973). The concept is a simple one and may be summarised as follows.

The factorial division of nutrients into components needed to match obligatory losses and those required for growth and activity has proved to be of great practical value in estimating both energy and protein requirements. This is shown schematically in Fig. 7.4. In the case of energy requirements, BMR represents the 'obligatory loss' of energy and is therefore used as the basis for building up the energy requirements. A similar approach may be taken for the estimation of protein requirements. Obligatory nitrogen loss (ONL) is defined as the inevitable loss of nitrogen in the urine, faeces and other routes in a subject fed a non-protein diet for 10–14 days (Fig. 7.5). After the first few days the nitrogen loss in the urine reaches a stable plateau and this point is taken as the obligatory urinary nitrogen loss (OUNL). Of the three routes of nitrogen loss, i.e. urine, faeces and integumental (skin, saliva, etc.), urinary nitrogen loss represents the largest component and its magnitude will finally dictate the total protein requirements (Table 7.6). Thus:

Obligatory nitrogen loss (ONL) = Obligatory urinary nitrogen loss (OUNL) + Faecal nitrogen loss (FNL) + Integumental nitrogen loss (INL)

Table 7.6. *Obligatory nitrogen losses in adult men on a protein-free diet*

Route	mg N per kg body weight	mg N per unit of basal energy	
		(kcal)	(kJ)
Urine	37	1.4	0.33
Faeces	12	0.4	0.10
Skin	3	0.13	0.03
Miscellaneous	2	0.08	0.02
Total	54	2.0	0.48

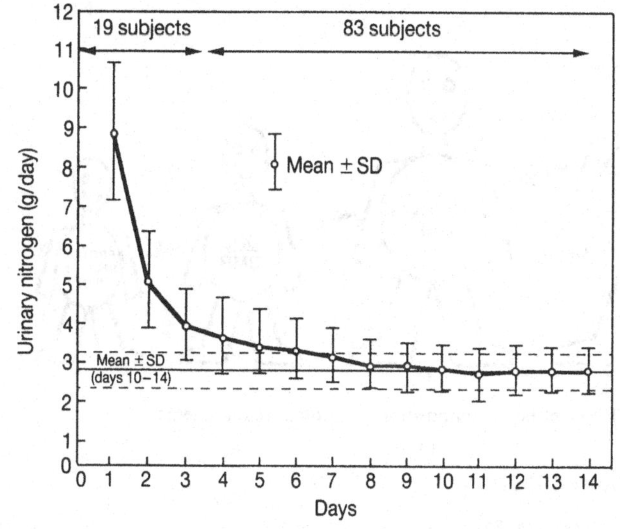

Fig. 7.5. Effect of a protein-free diet, adequate in energy, on urinary nitrogen excretion in humans.

or $$\mathrm{ONL} = \mathrm{OUNL} + \mathrm{FNL} + \mathrm{INL} \tag{7.8}$$

Since a subject's protein requirements are computed on the basis of the amount required to replace or match ONL, the greater OUNL, the greater will be an individual's protein requirements.

Having noted that ONL contributes significantly to the estimation of human protein requirements, we can next explore the relationship between FUNL and ONL. The former is the nitrogen loss during total energy and protein restriction (fasting) and the latter is the nitrogen loss during the feeding of a non-protein diet adequate in energy.

Several years ago Brody (1945), using allometry, showed ONL in a range of homeotherms to be of the form:

$$\mathrm{ONL} = 272W^{0.75} \tag{7.9}$$

Table 7.7. *Fasting urinary nitrogen loss (FUNL) and obligatory nitrogen loss (ONL) (mean ± SD) and the ratio FUNL/ONL in man*

Subjects	Age (years)	FUNL (mg/kg per day)	Age (years)	ONL (mg/kg per day)	FUNL/ONL
Infants	Newborn	78 ± 30	4–6 months	57	1.3
Young men	23.5 ± 3.7	156 ± 26	22.8 ± 0.6	89	1.7
Old men	82.1 ± 3.9	125 ± 14	70.13 ± 1.55	80′	1.5

From Henry *et al.* (1986).

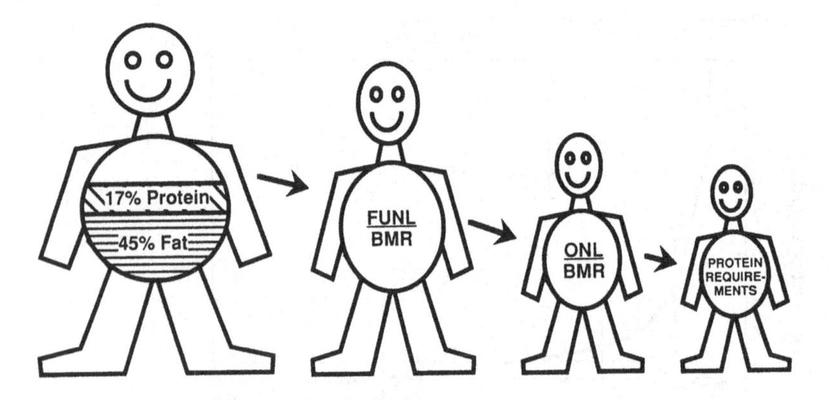

Fig. 7.6. Effect of body composition on protein requirements.

Henry (1984) showed that:

$$\text{FUNL} = 418\,W^{0.75} \tag{7.9}$$

Since the mass exponents are the same, using the allometric cancellation technique, dividing equation (7.6) by (7.9), we have:

$$\text{FUNL} = 1.5\ \text{ONL} \tag{7.10}$$

This is an important quantitative relationship. What it shows is that FUNL and ONL are tightly linked and that the processes of protein metabolism during starvation and during feeding of a non-protein diet may share some common metabolic pathways. The general validity of the relationship between FUNL and ONL shown in equation (7.10) may be tested using published values for ONL and FUNL in humans. This is shown in Table 7.7. Using published values for the initial FUNL and ONL in infants, young men and old men, after 2–3 days of fasting or on a protein-free diet, the FUNL/ONL ratio obtained (Table 7.7) is similar to that predicted using allometry (equation 7.10).

In an earlier section it was noted that adiposity, i.e. the amount of body

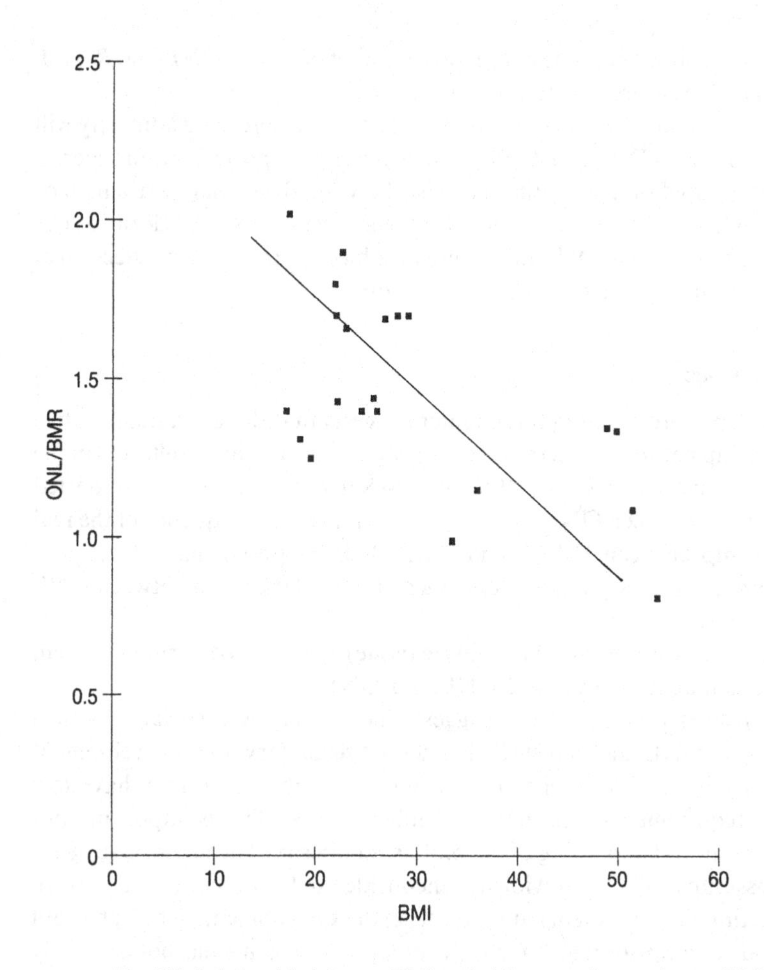

Fig. 7.7. Relationship between BMI and obligatory nitrogen loss to BMR ratio (ONL/BMR) in humans.

fat in an individual, significantly influenced protein mobilisation: i.e. the *greater* the adiposity, the *smaller* the FUNL. Now, linking the above relationship between FUNL and ONL, this should suggest a *smaller* ONL in those with a *smaller* FUNL. This is shown schematically in Fig. 7.6. Since ONL dictates an individual's protein requirement, subjects with a smaller ONL should have smaller protein requirements. One way we can test this is to examine the ONL/BMR ratio in human subjects fed a non-protein diet.

Whilst literature values where ONL, BMR values and body composition were measured in the *same* subject are limited, Fig. 7.7 shows a compilation of some values. In this figure, body mass index (BMI) has been used as a proxy for adiposity (Garrow, 1983). Given the limitations of the analysis, what is encouraging is the general trend of a *lower ONL/BMR* with increasing BMI (greater adiposity).

It may therefore be concluded that subjects with increasing adiposity will have a lower FUNL and ONL, and hence lower protein requirements. Since in a population a continuum exists between those that have a high or low FUNL/BMR ratio, this should lead to a varying ONL/BMR ratio. The observed variation in ONL in a group of subjects is therefore a reflection of their pattern of protein tissue mobilisation.

Conclusion

The evidence presented in this chapter suggests that there is a quantitative relationship between protein and energy metabolism. This is reflected in the association between FUNL, ONL and BMR in humans. The often-reported relationship between ONL and BMR is an indirect consequence of the real relationship between FUNL and BMR. It is proposed that FUNL and BMR are the primary parameters of which the relationship between ONL and BMR is a derivative.

Body composition (notably adipose tissue) appears to determine protein metabolism as measured by FUNL and ONL.

The results presented here suggest that obesity is associated with a lowering of ONL and that such changes are secondary to the development of obesity itself. The implication of this – that the obese may have low protein requirements – has not been fully explored. This is important, not only for our understanding of the biology of obesity. The analysis suggests the possibility that a previously unconsidered factor could lead to an underestimation of protein requirements in these experiments. The significant role body composition may play in protein requirements has not been fully explored. The evidence presented in this chapter suggests that body composition certainly has an important role to play in tissue protein breakdown and the determination of human protein requirements. It is hoped that this chapter will stimulate further work in this challenging area.

References

Benedict, F.G. (1915). *A Study of Prolonged Fasting*. Washington, D.C.: Carnegie Institute of Washington, publication no. 203.

Benedict, F.G. & Talbot, F. (1921). *Metabolism and Growth from Birth to Puberty*. Washington D.C.: Carnegie Institute of Washington, publication no. 302.

Behnke, A.R. (1953). The relation to lean body weight to metabolism and some consequent systematisations. *Annals of the New York Academy of Science*, **56**, 1095–42.

Benedict, F.G. & Fox, E.L. (1934). Energy metabolism in the wild rat. *American Journal of Physiology*, **108**, 285–91.

Brody, S. (1945). *Bioenergetics and Growth – with Special Reference to the Efficiency Complex in Domestic Animals*. New York: Reinhold.

Calloway, D.S. & Spector, H. (1954). Nitrogen balance as related to calorie intake and protein intake. *American Journal of Clinical Nutrition*, **2**, 405–10.

Elia, M. (1992). Energy expenditure in the whole body. In *Tissue Determinants Cellular Corollaries*, ed. J.M. Kinney, New York: Raven Press.

FAO (1973). Nutrition meeting report no. 52, WHO technical report series, no. 522, *Energy and Protein Requirements*. Geneva: World Health Organization.

Garrow, J.S. (1983). Indices of adiposity. *Nutrition Abstracts and Reviews*, **53**, 697–708.

Gilder, H., Cornell, G.N. & Thorbjarnarson, B. (1967). Human energy expenditure in starvation estimated by expired air analysis. *Journal of Applied Physiology*, **23**, 292–303.

Henry, C.J.K. (1984). Protein energy interrelations and the regulation of body composition. Ph.D. Thesis, University of London.

Henry, C.J.K., Rivers, J.P.W. & Payne, P.R. (1986). Does the pattern of tissue mobilisation dictate protein requirements? *Human Nutrition: Clinical Nutrition*, **40C**, 87–92.

Huxley, J.S. (1932). *Problems of Relative Growth*. London: Methuen.

Kleiber, M. (1932). Body size and metabolism. *Hilgardia*, **6**, 315–53.

Kral, J.G. & van Itallie, T.B. (1993). *Recent Developments in Body Composition Analysis*. London: Smith-Gordon.

Morgulis, S. (1923). *Fasting and Undernutrition*. New York: Dutton.

Payne, P.R. (1975). Safe protein: calorie ratio in diet. *American Journal of Clinical Nutrition*, **28**, 281–86.

Payne, P.R. & Dugdale, A.E. (1977). A model for prediction of energy balance and body weight. *Annals of Human Biology*, **4**, 525–35.

Richet, C.H. (1890). *Archives de Physiologie*, **2**, 483.

Rubner, M. (1883). Die Gesetze des energie verbrauchs bei der ernahrung. *Zeitschrift fuer Biologie*, **19**, 535.

Takahira, H. (1925). Metabolism during fasting and subsequent refeeding, vol. 1. Tokyo: Imperial Government Institute for Nutrition.

Van Noorden, C. (1907). *Metabolism and Practical Medicine*, vol. 1. London: Heinemann.

Von Dobeln, W. (1956). Human standard and maximal metabolic rate in relation to fat-free body mass. *Acta Physiologica Scandinavica*, **37**, 2–79.

8 *Prediction of adult body composition from infant and child measurements*

MARIE FRANÇOISE ROLLAND-CACHERA

Introduction

Anatomists demonstrated many years ago that organs grow at different rates, and that these rates can differ from the growth rate of the body as a whole (Forbes, 1978). Further, auxologists have shown that children grow at a variety of rates: they can play *lento* or *allegro* (Tanner, 1986). The first case corresponds to a normal process where growth is organised in successive steps, while in the second case individual variation due to genetic and/or environmental factors influences the growth process. This variation makes it difficult to predict adult body composition from childhood measurements. However, growth is affected by hormonal status, so that childhood is a good time to study the relationship between anthropometry and hormonal status, and to analyse the influence of environmental factors such as nutrition. In general, patterns of growth give more useful information than absolute levels of anthropometric measurements. Better understanding of factors influencing body composition can improve prediction of adult status and help to propose strategies for reducing the risk factors of various diseases.

Use of anthropometric measurements

Anthropometric measurements can be used in several ways: directly (e.g. skinfolds), as indices (e.g. weight/height^2, the Quetelet or body mass index (BMI)), areas (e.g. upper arm muscle area (UMA) based on arm skinfolds and arm circumference) or in regression equations relating body density to anthropometric measurements for a reference population. In addition, various ratios can be used to predict body shape and proportion.

Direct measurements and the BMI predict the level of fatness, while UMA and the regression equations predict body composition (i.e. fat mass (FM), fat-free mass (FFM) and % body fat (%BF)). Fatness level as predicted from direct measurements, BMI or UMA, does not depend on a reference population, whereas a regression equation does. These methods have been reviewed previously (Forbes, 1978; Rolland-Cachera, 1993).

Description of some indicators

Height and weight

Height is a useful indicator of nutritional status; undernutrition delays height growth, while overnutrition accelerates it (Forbes, 1977). Reduced stature, or 'stunting' (Waterlow & Rutishauser, 1974), and increased stature, or 'bolting' (Rolland-Cachera *et al.*, 1991*a*), can both be interpreted as adaptations to inappropriate energy intake, limiting under- and overweight. If not adjusted for height, weight gives limited information on body composition, and is useful only when followed longitudinally.

Skinfold thickness (SF)

Skinfold thickness is a measure of fatness. Subcutaneous fat constitutes a large part of total body fat, but its proportion varies with age, sex and degree of adiposity. In the newborn, the percentage of subcutaneous fat is higher than later in life. The different sites correlate differently with total-body fat and with %BF (Roche *et al.*, 1981): the triceps SF (Tri) has a better correlation with %BF while the subscapular SF (SS) correlates better with total-body fat.

Body mass index (BMI)

The BMI has been found to be associated with body composition (Keys *et al.*, 1972; Garrow & Webster, 1985) and nutritional status. It also has a high correlation with body fatness and a low correlation with height (Rolland-Cachera *et al.*, 1982). In addition, it is the only index of the form weight/height^n that changes during childhood in the same way as SF (Fig. 8.1).

Body fat distribution (BFD)

The influence of body fat distribution on health, first suggested by Vague (1956), is a topic of current interest. Abdominal (android) obesity has been found to be associated with an increased risk of glucose intolerance, coronary artery disease and stroke in both men and women. Locations of fat and fat patterns are assessed in adults by comparing central and peripheral measurements, usually as SF (e.g. trunk SF/extremity SF ratio). A high value corresponds to an android pattern, and a low value to a gynoid pattern. Compared with the waist/hip circumference ratio (the

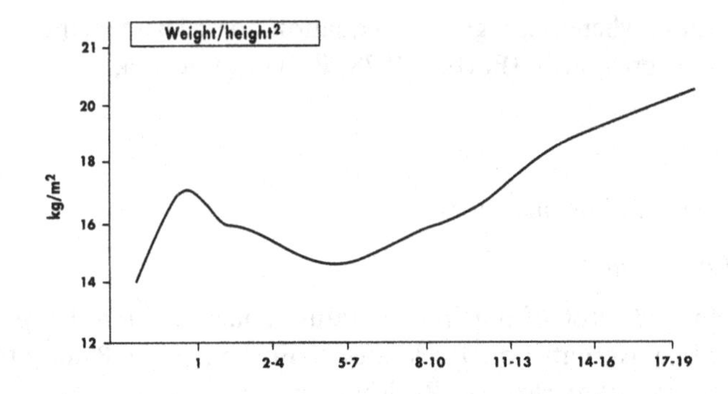

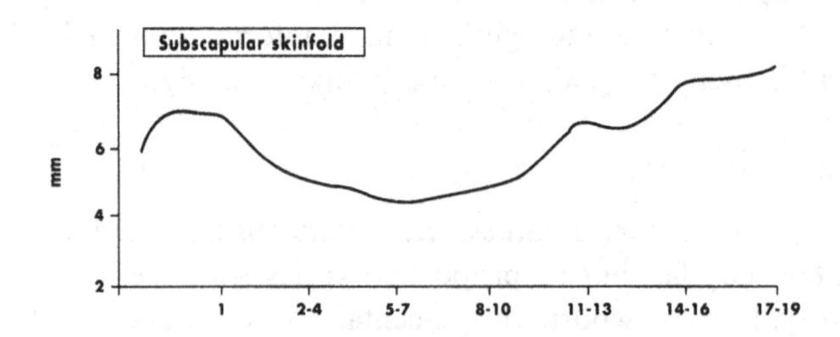

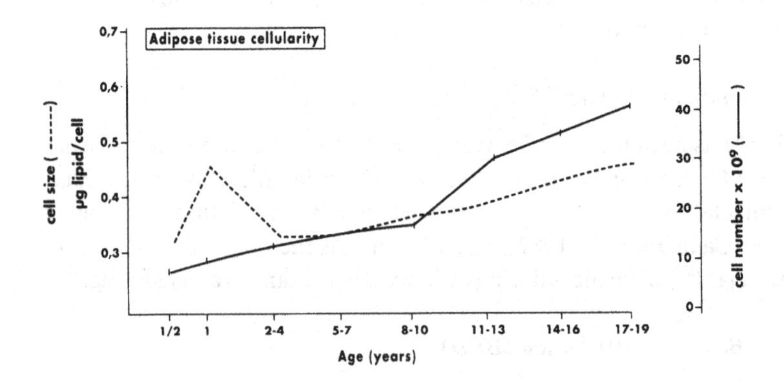

Fig. 8.1. Development of fatness assessed by body mass index (BMI) and skinfolds, and corresponding changes in adipose tissue cellularity. From Rolland-Cachera *et al.* (1982), Sempé *et al.* (1979), Knittle *et al.* (1979).

other method for assessing BFD), the trunk/extremity SF ratio has been found to be a better indicator of fat pattern in children (Sangi & Mueller, 1991).

Body proportions

Statural body proportions (SBP) are assessed either by the upper-segment length/stature ratio or upper/lower segment ratio. A low value (i.e. relatively long legs) corresponds to an android type, and a high value to a gynoid type.

Frame

Biacromial/biiliac diameter ratio is a characteristic of the frame. A high ratio corresponds to the android type and a low ratio to the gynoid.

Selection of methods for different uses

For nutrition surveys

The selection of appropriate measures for nutrition studies depends on several criteria, and Table 8.1 lists the limitations of the various methods. BMI cannot distinguish between fat and fat-free mass, but it does measure internal fat, while SF measures subcutaneous or 'low-risk' fat. For these reasons, SF is better for measuring body composition (%BF), while BMI is good for assessing the relationship between internal fat and risk factors. BMI has other advantages over SF: child–adult correlations are better than for SF (Rolland-Cachera *et al.*, 1989) and weight and height are more widely available than SF. Of the various sites for measuring SF, trunk sites (e.g. SS) are better than extremity sites (e.g. TRI) as they correlate better with internal fat. TRI is a better measure of body composition, while BMI and SS are better for measuring nutritional status and risk factors such as obesity. SF combined with BMI distinguishes between the overweight obese and overweight non-obese (e.g. athletes). Combined with arm circumference, TRI assesses %BF from the areas formula (UMA), and combined with SS it assesses BFD (trunk/extremity SF ratio). In summary, all the measurements have useful properties: BMI on its own, but more particularly combined with SS and TRI SF and arm circumference, when it provides extra information on body composition.

For the clinician

Weight and height reference charts are important for monitoring height gain and weight gain, but they do not assess nutritional status well. During growth, weight increases with both age and height, and these associations

Table 8.1. *The limitations of different methods in nutrition studies*

Characteristics	Age	Males	Females
Relationship with body composition			
% BODY FAT (density) (Roche *et al.*, 1981)			
Choice of SF and indices	6–13 years	TRI	TRI
	13–18 years	TRI	TRI
TOTAL BODY FAT (density) (Roche *et al.*, 1981)			
Choice of SF and indices	6–13 years	BMI	BMI
	13–18 years	SS	BMI
TOTAL BODY FAT (CT Scan) (Seidell, 1987)			
Choice of SF and indices	Adults	BMI	BMI
INTERNAL FAT (MRI, CT Scans)			
Choice of SF (9 sites) (Fox *et al.*, 1993	11.5 years	None	SS
Choice of indices and SF (Seidell *et al.*, 1987)	Adults	BMI	BMI
COMMUNALITY			
Choice of SF (10 sites) (Siervogel *et al.*, 1982)	11–16 years	All	All
Response to nutritional intervention			
Choice of SF (Himes, 1988)	Child/adult	Trunk SF	Trunk SF
Specificity (Himes & Bouchard, 1989)			
Choice of regression equation, SF, indices	8–19 years	TRI	BMI
Relationship with risk factors			
Choice of indices and SF (Sangi & Mueller, 1991)	12–17 years	BMI	BMI
Choice of SF (Freedman *et al.*, 1987)	6–18 years	Trunk SF	Trunk SF
Choice of SF (Ducimetière *et al.*, 1986)	Adults	–	Trunk SF
Choice of SF (Lapidus *et al.*, 1984)	Adults	Trunk SF	–
Tracking (Rolland-Cachera, 1989)			
Choice of indices and SF	Child/adult	BMI	BMI
Choice of SF		Trunk SF	Extr SF

SF, skinfolds; TRI, triceps SF; SS, subscapular SF; BMI, body mass index; MRI, magnetic resonance imaging; CT, computed tomography.
(–), no data.

reflect changes in stature rather than fatness. Conversely, changes in BMI reflect changes in body fatness. The BMI reference chart rises during the first year of life, so that by the age of 1 year children look chubby. They then slim down, and by 6 years children on the 50th centile look thin. Equally, children on the 10th centile look severely underweight, and those on the 90th centile look normal or slightly overweight. Consequently, most obesity at 6 years is not detected, and sometimes children who look thin are

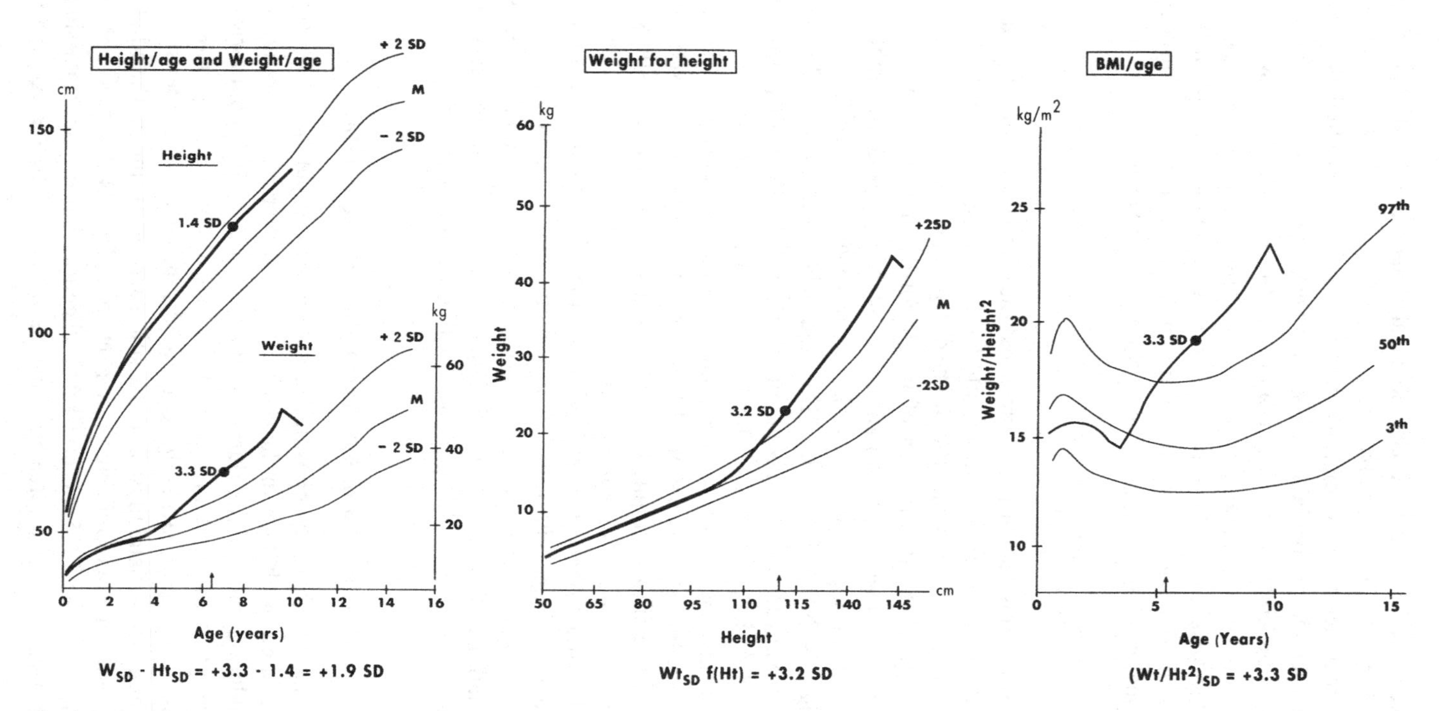

Fig. 8.2. Comparison of methods for assessing weight development: (1) weight- and height-for-age; (2) weight-for-height; (3) weight/height^2 for age.

forced to eat in order to reach their 'ideal' weight. Beyond the age of 6 years, the BMI increases steadily.

For clinicians, the choice of method to assess overweight depends on the reference charts available. Weight-for-height charts are convenient, but ignore the child's age. Weight- and height-for-age charts are widely used, and it is common practice to assess nutritional status by comparing the child's weight and height centiles. If the two centiles, or equivalently the standard deviation scores (SDS), are similar, the child is often considered as normal weight-for-height. This is equivalent to using the formula:

$$\mathrm{Wt_{SDS}} - \mathrm{Ht_{SDS}} \tag{8.1}$$

However, this calculation assumes that the correlation between weight and height is perfect (i.e. = 1), when in fact it is about 0.7 until puberty, and declines thereafter (Cole, 1986). The distribution of weight-for-age is much wider than the distribution of weight-for-height, so that the degree of under- and overweight in light or heavy children is considerably underestimated using equation (8.1). For example, in a girl aged 6.5 years whose weight is 29 kg (i.e. + 3.2 SDS) and height 120 cm (+ 1.3 SDS), her overweight as assessed by equation (8.1) is + 1.9 SDS; by weight-for-height is + 3.2 SDS; and by BMI for age is + 3.3 SDS (Fig. 8.2).

The last two methods give similar results, and show that the child is obese, while using equation (8.1) her overweight is only 1.9 SDS. Since it takes age into account, the BMI method improves the estimation of weight-for-height at adolescence. In addition, weight changes appear earlier and more clearly on BMI charts than they do on weight-for-age or weight-for-height charts and, finally, monitoring the individual BMI curve on a standard chart distinguishes between permanent and transient weight deviations. This avoids restricting the child's intake unnecessarily or, conversely, delaying dietary intervention when it is needed.

Development of body composition during growth

Changes in body composition during growth occur in clearly separate stages. Large changes in the chemical composition of FFM occur in early childhood, particularly during fetal life. Moulton (1923) put forward the concept of chemical maturity: that at some point during growth body composition approaches that of the adult, and hence is considered 'mature'. In humans, skeletal muscle composition achieves values in the adult range for potassium, chloride and water by about 6 years (Nichols *et al.*, 1968). Cureton *et al.* (1975) observed that 'chemical maturity for potassium and the density of the fat-free body has been achieved by the age of 8 years'.

However, chemical maturity occurs at different stages of growth for the various constituents of the body.

Anthropometric development reflects these differences: head circumference, which partly reflects brain development, reaches 85% of its adult value by the age of 1 year, while stature is less than 50% of the adult value (Sempé, 1988). This is consistent with the fact that bone mineral maturation occurs only in the third decade (Lohman *et al.*, 1984). Growth velocity varies from tissue to tissue, and can even move in opposite directions: height velocity, for example, is at a maximum when triceps velocity is at its nadir (Forbes, 1978). Low tracking (see below) of anthropometry, reflecting periods of rapid change and maturation, is observed at different periods in the different tissues: during early life (up to 6 years) for fatness, and at puberty for stature (Rolland-Cachera, 1993).

These chemical and anthropometric timing differences show that growth occurs progressively, in different stages, one tissue after another.

Anthropometric changes

Fat and fat-free mass changes

Mean changes. Fat-free mass and body fat change throughout life. During the early years, the proportion of body fat changes substantially (Fomon *et al.*, 1982), increasing during the first year, then decreasing and increasing again after 6 years. Fat-free mass increases steeply until 19 years in boys and 15 years in girls. Conversely, body fat increases until 17 years in girls, but only until 15 years in boys after which it decreases again (Forbes, 1972). Consequently %BF increases up to age 17 years in girls, but decreases steeply after 13 years in boys.

Individual changes. Most anthropometry increases during adolescence, so that absolute changes are of less concern than shifts across centiles. Tracking in a measurement implies maintenance of the same rank order among a group of individuals as they age. Tracking of fatness is well documented; Poskitt & Cole (1977) showed that most fat babies do not stay fat, while Abraham *et al.* (1970) showed that by the age of 9 years the weight pattern for later life has largely been set. Using the BMI, Cronk *et al.* (1982) observed that intra-individual variation appears to be greater during infancy and adolescence than during childhood, and that the degree of continuity is greater between childhood and adolescence than between infancy and childhood. Similarly Rolland-Cachera *et al.* (1987) observed that most fat children do not stay fat. However, the risk of being fat as an adult is doubled in fat infants compared with their lean peers.

Changes of tracking with age can be seen from the BMI curve (Fig. 8.3).

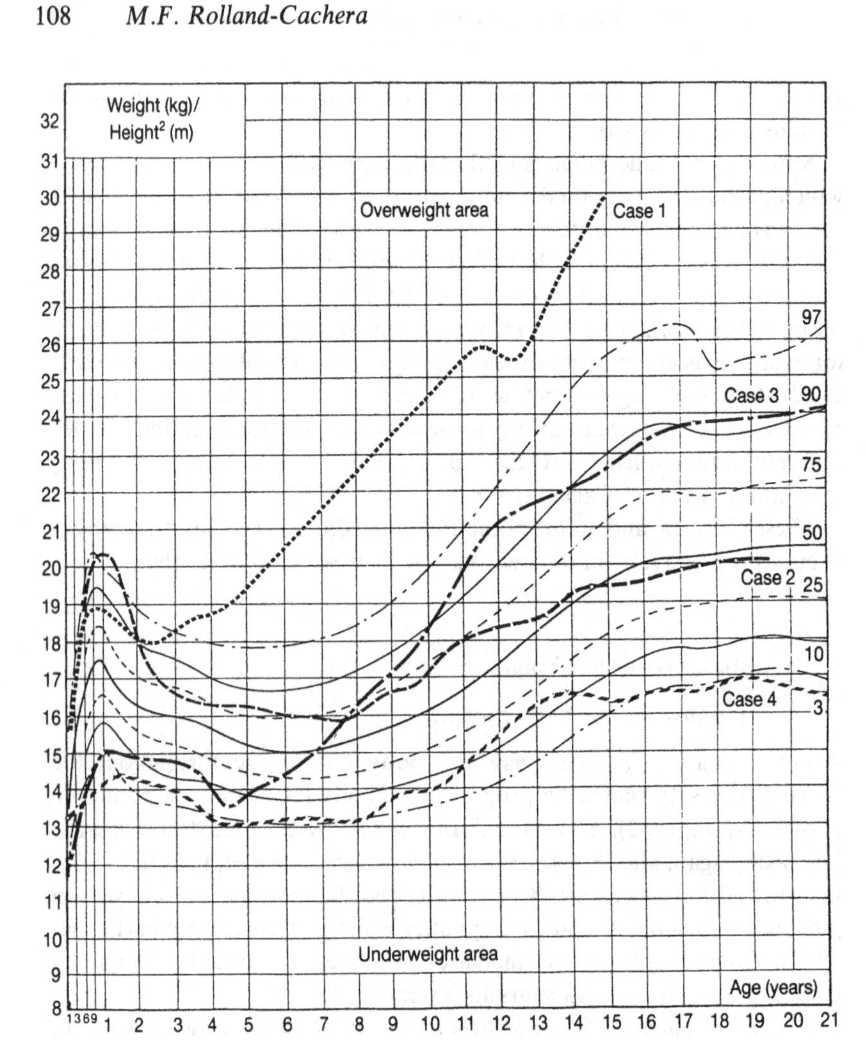

Fig. 8.3. Four examples of BMI development: case 1, fat at 1 year, stayed fat after early rebound at 2 years; case 2, fat at 1 year, became lean after late rebound at 8 years; case 3, lean at 1 year, became fat after early rebound at 4.5 years; case 4, lean at 1 year, stayed lean after late rebound at 8 years. After Rolland-Cachera *et al.* (1987).

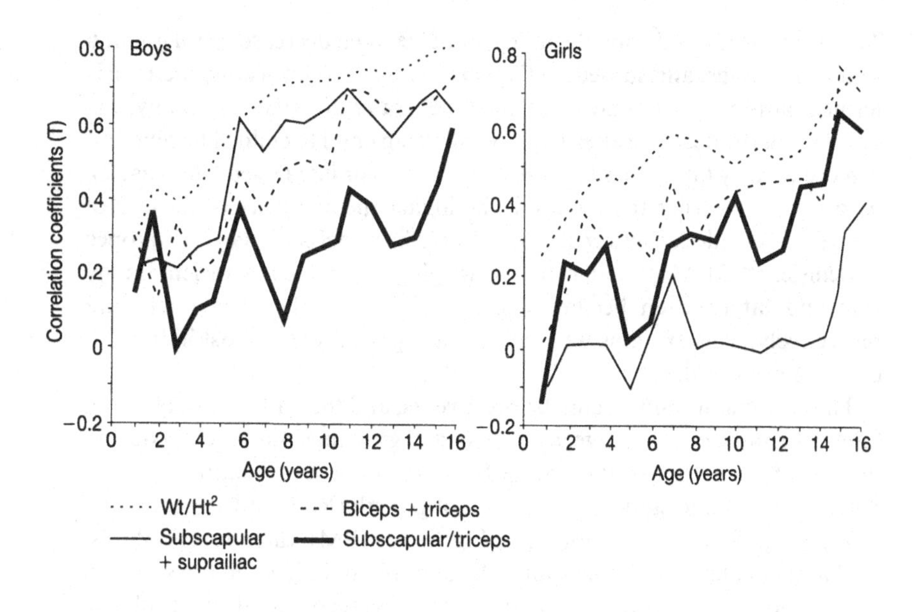

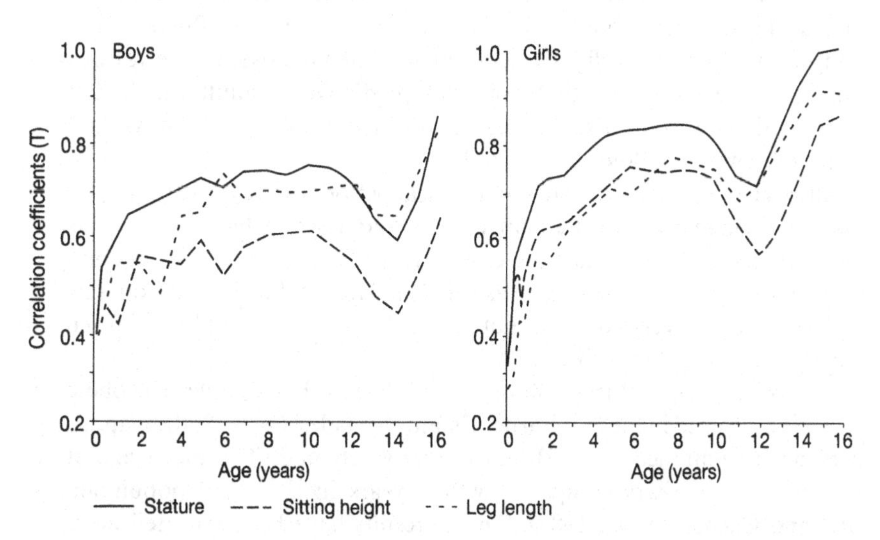

Fig. 8.4. Child–adult correlations for fatness level and distribution (Rolland-Cachera *et al.*, 1989; 1990) and for stature (Sempé *et al.*, 1979).

The curve increases during the first year of life, then decreases until about 6 years on average, and subsequently increases again. This second increase is termed 'adiposity rebound' (Rolland-Cachera *et al.*, 1984). Actually, the duration of the decrease after 1 year varies from child to child. The rebound can occur early (at 2 years for case 1 in Fig. 8.3) or late (8 years for case 2). As a rule, the earlier the rebound, the higher the adiposity at the end of growth (Rolland-Cachera *et al.*, 1984, 1987; Siervogel *et al.*, 1991; Prokopec & Bellisle, 1993). Most under- or overweight early in life is transient; many lean and fat children become average after an early or late rebound respectively. This explains why, before the age of 8 years, most individual curves cross centiles.

The rebound usually occurs before 8 years, and thereafter most children follow the same centile range until the end of growth. Consequently during infancy, child–adult correlations for BMI are low. They rise until the age of 8 years and remain stable until the end of growth (Rolland-Cachera *et al.*, 1989) (Fig. 8.4). Correlations are higher for BMI than for skinfolds, probably because FFM is less variable than BF, and higher in boys than in girls. In addition, prediction is better for fatness recorded at the trunk site than at the arm site in boys, while the reverse is true in girls. Prediction of body fat distribution is slightly better in girls than in boys. Of the various skinfolds, the indicators that are both predictive of adult status and associated with adult risk factors are the trunk skinfold in boys and trunk/extremity skinfold ratio in girls.

Adolescence is often considered a critical period for the development of obesity, yet in many cases the changes seen at adolescence have their origins many years before. For example case 3 (Fig. 8.3) is a girl whose rebound occurred at 4.5 years, when she was still lean. Her BMI increased from this time on, but her overweight was not detected until the age of 11 years when her BMI reached the 90th centile.

Almost all obese children rebound before 6 years. In a sample of 62 obese subjects examined in the Hôpital des Enfants Malades, Paris (R. Rappaport, personal communication), 60 had an early rebound. The mean age at rebound was 3 years compared with 6 years in a normal population (Rolland-Cachera *et al.*, 1987). Similar results have been reported more recently (Girardet *et al.*, 1993). Age at rebound is a good predictor of subsequent risk. BMI monitoring is particularly useful for identifying the lean who are at risk of becoming overweight, and for distinguishing between the fat who will stay fat and the fat who will slim down.

The mean BMI pattern reflects fatness and adipocyte cellularity development (Knittle *et al.*, 1979). Cell size increases during the first year of life and then decreases, while cell number increases slowly during early life and rapidly after 8 years (Fig. 8.1). These changes suggest that the BMI

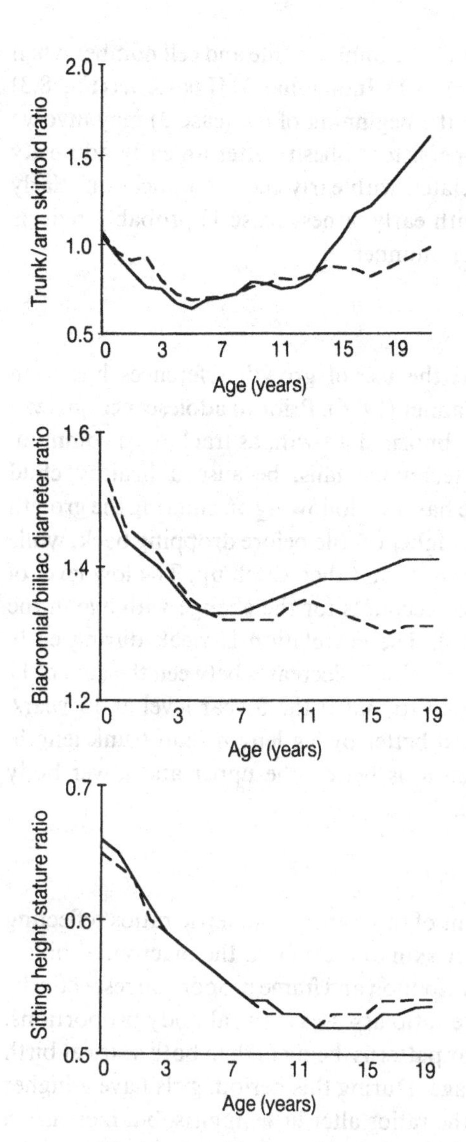

Fig. 8.5. Development of body fat distribution (trunk/extremity (arm) skinfold ratio), frame proportions (biacromial/biiliac diameter ratio) and statural body proportions (upper-segment (sitting height)/stature) in boys (continuous line) and girls (dashed line). After Rolland-Cachera *et al.* (1990), Sempé *et al.* (1979).

pattern reflects mainly cell size at the beginning of life and cell number when fatness starts to increase again (Fig. 8.1). Individual BMI patterns (Fig. 8.3) suggest that transient obesity at the beginning of life (case 2) may involve increased adipocyte size, while persistent obesity after an early adiposity rebound (case 3) could be associated with early cell multiplication. Early adiposity rebound associated with early fatness (case 1) probably reflects both large cell size and large cell number.

Stature

Prediction of adult stature and the use of growth references has been comprehensively described by Tanner (1986). Prior to adolescence, growth charts are useful for identifying abnormal growth, as tracking is common. In adolescence, however, the technique fails, because a healthy child departs from the centile he or she has been following on entering the growth spurt. Early maturers move to a higher centile before dropping back, while late maturers move to a lower centile and then catch up. The low level of tracking observed at adolescence accounts for the change with age in the child–adult correlation (Fig. 8.4). The correlation is weak during early childhood, but rises steeply until 5 years. It decreases between the ages of 11 and 14 years, and then rises again to match the 5 year level at 15 years. Adult stature in boys is predicted better by leg length than trunk length, while in girls, where the prediction is better, the upper and lower body segments are equally predictive.

Ratios reflecting body shape

Fig. 8.5 presents the development of three anthropometric ratios reflecting body shape. The trunk/extremity skinfold ratio and the biacromial/biiliac diameter ratio assess body fat distribution and frame proportion respectively, while the upper-segment/stature ratio assesses statural body proportions. The first two ratios follow similar patterns, being high in both sexes at birth and decreasing until 6 years of age. During this period, girls have a higher ratio than boys. Subsequently the ratios alter little in girls, but increase in boys due to decreased fatness at the trunk site and increased biacromial diameter.

The upper-segment/stature ratio is high in both sexes at birth and decreases until adolescence. During infancy girls have a slightly lower ratio than boys. Subsequently, the ratio increases in both sexes but earlier in girls than in boys (11 years as against 13), and the girls' adult ratio is also higher. Taller adults tend to have a lower upper-segment/stature ratio, i.e. a more android shape (personal observation).

Newborns of both sexes display male characteristics for body fat distribution and frame proportion, and female characteristics for statural body proportion. In all three cases, girls at birth display more android characteristics than boys, with more fat at the trunk site, shorter biiliac diameter, and shorter trunk length. Conversely, as adults, females have more fat at the arm site, shorter biacromial diameter and shorter leg length.

The male characteristics in girls and female characteristics in boys seen at birth stress that anthropometric indicators at this age have very little predictive value for adult status. This is consistent with the low infant–adult correlations that are seen for most measurements. The correlations are even negative for some measurements, such as subscapular/tricep skinfold ratio in girls (Fig. 8.4).

Hormonal status during growth

Growth is the result of a complex interaction between nutrition, anabolic and catabolic factors, and the response of the target organs. Hormonal status and body composition are closely related through direct or metabolic influences. Hormone levels in the circulation and their regulation alter markedly with age and sex. As hormones are secreted at clearly distinct stages, growth is useful for examining the possible relationships between hormonal status and body composition.

Role of hormones in various tissues

During growth, hormones such as insulin and growth hormone (GH) have a critical role. GH secreted by the pituitary gland acts not only on longitudinal bone growth but it also has anabolic, lipolytic and antinatriuretic effects. GH exerts its growth-promoting effects by stimulating the synthesis of insulin-like growth factors (IGFs) in the liver and other tissues.

Whenever growth and its regulation is discussed, the term 'somatomedin' comes into play (Zapf *et al.*, 1982). The terms 'somatomedin' and 'insulin-like growth factor' (IGF) are synonymous for one particular class of peptides. Two human somatomedins, IGF1 and IGF2, have been isolated, and both possess anabolic and mitogenic properties. They are GH-dependent, but this does not exclude other direct action by GH. The basis for the biological properties of IGF1 and IGF2 rests in their structural analogies with insulin. They elicit the same qualitative biological responses as insulin. These responses are mediated by interaction with either insulin receptors or specific IGF receptors, depending on the tissue. Consequently the biological potencies of IGF1 and IGF2 differ from that of insulin, and the potency ratios between insulin and the IGFs vary considerably from one tissue to another. These ratios give some clue as to the physiological role of IGFs compared with insulin.

The somatomedins are present throughout the nervous system. They probably regulate the growth of the developing nervous system and act as maintenance hormones by their anabolic action on cell metabolism in the mature nervous system (Sara *et al.*, 1981). Only IGF2-related and not IGF1-related peptides are present in the cerebrospinal fluid. The somatomedins appear to be endogenously produced by neural tissue (Binoux *et al.*, 1981). IGF2 seems to result from central nervous system production rather than transport from peripheral circulation across the blood–brain barrier (Hall and Sara, 1984).

The biological effects of IGFs in various tissues have been described by Zapf *et al.* (1984). In adipose tissue or isolated fat cells, IGF1 and IGF2 stimulate glucose transport, in turn stimulating glucose oxidation and lipid synthesis from glucose. Compared with insulin, IGF1 and IGF2 are only one-fiftieth and one-hundredth as potent, respectively, in enhancing glucose metabolism in the fat cell. The potency difference between IGF1 and IGF2 in isolated fat cells is consistent with their potency difference in competing for the insulin receptor of the adipocyte. *In vivo*, IGFs act like insulin to reduce blood sugar, and IGF1 is somewhat more effective than IGF2. Insulin and the IGFs differ greatly in their effect on lipid synthesis in adipose tissue; insulin causes a tenfold stimulation of [^{14}C]glucose incorporation into adipose tissue total lipids, whereas IGF1 and IGF2 have only a very minor effect. Consequently, despite their antilipolytic action, the competition for insulin receptors and the lower potency on lipid synthesis probably cause IGFs to have a paradoxically lipolytic effect.

In the heart and skeletal muscle IGF1, like insulin, stimulates the glucose transport system. The biological potency of IGF compared with insulin is 1:2 to 1:5 in the perfused rat heart and 1:10 to 1:20 in the striated muscle. No significant differences in potency are found between IGF1 and IGF2. In osteoblast-like cells, IGF1 stimulates RNA and protein and glycogen synthesis, with a potency 5–7 times that of IGF2 and approximately 50 times that of insulin. So in contrast to the situation in adipose tissue, in osteoblasts, IGFs are more potent than insulin. These differences in potency suggest that the different tissues develop successively, under the same hormonal influence.

Sex and age variation

Fetal insulin and IGFs emerge as the principal endocrine regulators of prenatal growth (Brauner, 1993). IGFs increase during gestation, but decline during the final weeks (IGF1 more than IGF2) (Girard, 1989). In the newborn, GH is high (Cornblatt *et al.*, 1965). It falls rapidly during the first 2 weeks after birth and then more slowly to the end of the first year,

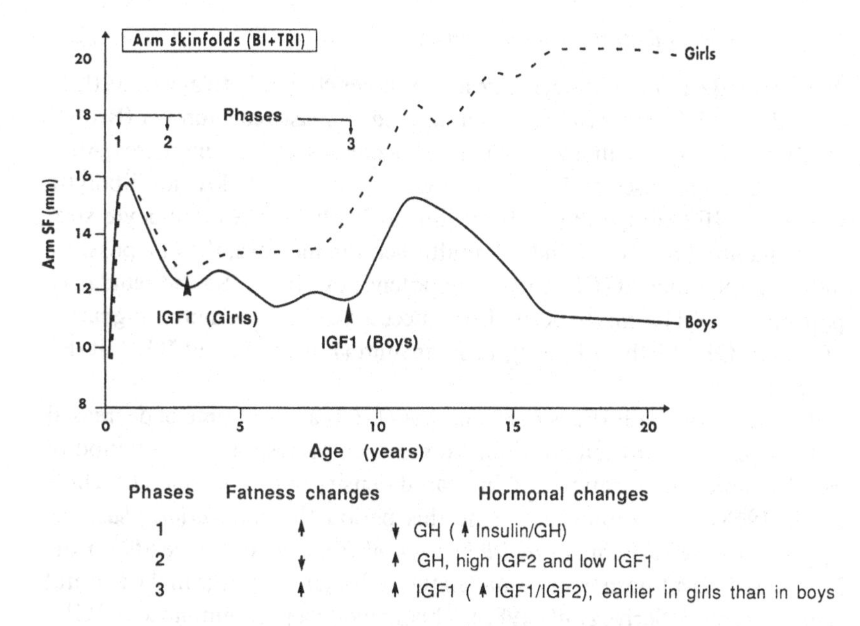

Phases	Fatness changes	Hormonal changes
1	↑	↓ GH (↑ Insulin/GH)
2	↓	↑ GH, high IGF2 and low IGF1
3	↑	↑ IGF1 (↑ IGF1/IGF2), earlier in girls than in boys

Fig. 8.6. Development of arm skinfold thickness (Sempé *et al.*, 1979) and a hypothesis about its relationship with hormonal status.

while the IGFs increase, IGF2 faster than IGF1. IGF2 reaches its adult value by 1 year of age, while IGF1 takes until age 8 years (Zapf *et al.*, 1981). Sex differences are seen in both concentration and timing: girls (7–18 years) have higher values of GH (Costin *et al.*, 1989) and a higher IGF1/IGF2 ratio than boys (Rosenfeld *et al.*, 1986). The steep rise in IGF1 occurs earlier in girls (3–5 years) than in boys (9–11 years) (Bala *et al.*, 1981; Rosenfeld *et al.*, 1986), and boys only catch up with the values in girls after age 15 years. Adult IGF1 declines while IGF2 remains high (Hall & Sara, 1984). The circulating levels of hormones represent a balance between production and utilisation (Hall & Sara, 1984). However, sex differences during growth may reflect differences in production.

Relationship between hormonal status and body composition

Body composition changes result from a balance between lipogenesis and lipolysis, which is probably regulated by a balance between insulin/GH, insulin/IGFs and IGF1/IGF2. The similarities in timing of the hormonal, cellular and anthropometric changes that occur during growth suggest a close relationship between these various components (Fig. 8.6).

Hormonal status and cellularity

As described earlier, adipocyte size and number change with age (Knittle *et al.*, 1979). GH decreases adipose cell size and increases cell number (Brook, 1973), and during the first year when GH decreases, cell size increases. After 1 year, GH increases and IGF2 is high. Because of its low antilipolytic potency, IGF2 competing with insulin probably reduces adipocyte size. Subsequently both IGF1 and cell multiplication increase, the one possibly inducing the other, IGF1 being more potent than IGF2. So the cellularity pattern (Fig. 8.1) seems to derive from successive phases of increasing ratios of insulin/GH (birth to 1 year), IGFs/insulin (1–6 years) and IGF1/IGF2 (after 6 years).

The period when IGF2 is predominant (1–6 years) is a time of decreased cell size and reduced cell multiplication, which corresponds to a period of rapid chemical maturation and increased density of fat-free mass (Nichols *et al.*, 1968). We propose to name this period the *maturation phase*, as opposed to the subsequent *multiplication phase*. Cell size depends on the location of the adipose tissue: small in the abdomen and large in the femoral region (Rebuffé-Scrive *et al.*, 1985). This suggests a predominance of IGF2 and IGF1, respectively, in these two sites. Also, the cellularity and metabolism of the mammary adipose tissue is more like abdominal than femoral adipose tissue (Rebuffé-Scrive, 1986), suggesting a higher IGF2 level at this site.

Various studies have shown that changes in adipocyte cellularity may extend to other tissues. Increased cell size and number in several organs and tissues have been reported in adult juvenile-onset obesity (Naeye & Roode, 1970), and increased fat-free mass is related to early-onset obesity in childhood (Knittle *et al.*, 1979). In addition, the islets of Langerhans in the pancreas are hyperplasic in the obese (Jeanrenaud, 1979). This similarity in the cellularity of the various tissues can be explained by a common hormonal status.

Hormonal status and fatness

Fetal fat deposition during the last trimester of pregnancy coincides with falling IGF levels (Girard, 1989), possibly related to a high insulin/IGF ratio. After birth, fatness changes correspond to hormonal changes (Fig. 8.6). The early adiposity rebound of the arm skinfold curve of girls compared with boys (3 versus 6 years) may be related to the girls' earlier IGF1 increase, and also to the increased lipoprotein lipase activity in the arm (Smith *et al.*, 1979; Rebuffé-Scrive *et al.*, 1985). The association between gynoid pattern and IGF1 is consistent with girls having a higher

IGF1 level than boys. In addition, IGF1 is negatively associated with visceral fat mass in men (Marin *et al.*, 1993). GH deficiency is generally reported in obesity (Pertzelan *et al.*, 1986), and the severity of the deficiency is related to the degree of overweight (Veldhuis *et al.*, 1991).

Hormonal status and body shape and proportions

As discussed above (Fig. 8.5), newborns have respectively male and female characteristics for body fat distribution and statural body proportions; these reflect hormonal status, and could correspond to low IGF levels. In the newborn, the proportions of IGF1 and IGF2 associated with binding proteins are very low, like those in patients with hypopituitarism (Lassare *et al.*, 1991). From birth to about 10 years of age, the android form of BFD becomes more gynoid and the gynoid type of statural body proportion becomes more android, probably as a consequence of increased IGFs. Site-specific changes in adipose tissue during childhood are similar to those observed with GH administration in GH-deficient children, i.e. a more peripheral distribution (Rosenbaum *et al.*, 1989). Successive low GH/insulin and IGF1/IGF2 ratios are likely to be responsible for height velocity decreasing in early life. The subsequent increase in height velocity during adolescence may correspond to a high IGF1/IGF2 ratio.

Growth indices and risk factors

Anthropometric measurements reflect body composition and hormonal status, and can be used as predictors of risk factors associated with hormonal status.

The four patterns of BMI development shown in Fig. 8.3 may reflect different hormonal states. The early fatness of case 1 could be attributed to low GH (and IGFs) associated with a high IGF1/IGF2 ratio, increasing the number of large adipocytes. The transient obesity in case 2 may be induced by delayed IGF production but normal GH, maintaining IGF2 release over an extended period (hence the late adiposity rebound). Case 3 probably corresponds to high IGFs (starting with high IGF2), steeply decreasing adipocyte size, and a subsequent early IGF1 increase causing early adipocyte multiplication and accelerated growth. Case 4 could be associated with low GH, but a high GH/insulin ratio and low IGF1/IGF2 levels.

The 'adiposity rebound' on the arm skinfold curve (Fig. 8.6) occurs earlier in girls than in boys. This is a sign of larger subcutaneous fat depots in girls. It may correspond to the earlier IGF1 release and could reflect the sex difference in metabolic risk associated with fat distribution. IGF1 tends to be low in boys during growth, but it may also vary genetically within as well as between the sexes. Indeed there is a high genetic determinant in body fat distribution (Ramirez, 1993). IGF1 is more variable than IGF2 under nutritional influence (Merimée *et al.*, 1982), and higher IGF1 in females could be better adapted to face metabolic stress during pregnancy. This would help cope with adverse environmental conditions such as caloric excess, and in the absence of pregnancy could provide a relief system. In subjects whose IGF1 production is low, caloric excess cannot be stored in the peripheral fat depots, so it can increase IGF2 and promote abnormal cell proliferation or various metabolic disorders. The IGF1 sex difference may account for the lower association between adolescent fatness and adult morbidity in girls than in boys (Must *et al.*, 1992).

The concept of better adaptation in females is consistent with the changes in lipoprotein lipase activity (LPLA) seen during pregnancy, which could be related to IGF1 level. LPLA is higher at the femoral than the abdominal site, and increases during the first 10 weeks of pregnancy. It decreases again during the latter part of pregnancy and lactation, and later at the menopause (Rebuffé-Scrive, 1985, 1986). IGF1 deficiency is associated with android fat pattern and metabolic disorders such as cardiovascular disease and non-insulin-dependent diabetes (Marin *et al.*, 1993). An increased incidence of cancer is found in lean subjects with centralised body fat mass (Lapidus *et al.*, 1988; Filipovsky *et al.*, 1993), which is also characterised by IGF1 deficiency. Abnormal IGF2 production by malignant tumours has been observed (Marquardt *et al.*, 1981), which could result from IGF1 deficiency and/or IGF2 excess promoting abnormal cell proliferation.

In addition to subcutaneous fat, other relief systems may exist. Hair also reflects hormonal status. Hippocrates (Joly, 1978) considered the brain as a gland, and hairs for the brain, like hairs for the other glands, 'collect the excess arising from fluxions and release it outside'. A recent study has shown an association between myocardial infarction and degree of baldness (Lesko *et al.*, 1993). Baldness is rare in females, and soon after birth when IGF demand is high, the newborn loses his or her hair. This may reflect high IGF production and utilisation respectively. Menstruation may also depend on the balance between hormonal production and utilisation. It starts when IGF1 (or IGF1/IGF2 ratio) is high (menarche), and disappears when production is low (malnutrition, intensive exercise (Smith *et al.*, 1987) or menopause) or when utilisation is high (pregnancy, lactation).

The greater variability of IGF1 status in girls is probably an advantage for metabolic adaptation. However, IGF1 is more variable than IGF2, which may also be detrimental in some situations, perhaps explaining higher weight fluctuations and a higher prevalence of anorexia and bulimia in girls (Story *et al.*, 1991). This suggests that other sex differences in mental disorder, such as more aggression in males (Hutt, 1972) or the predominance of psychosis in boys and neurosis in girls (Henderson, 1968), could be explained by these hormonal differences. Genetically obese male rats are more prone than females to stress (Guillaume-Gentil *et al.*, 1990). IGFs are involved in psychiatric disorders: for example cerebrospinal fluid levels of IGF2 are raised in Alzheimer's disease (Sara *et al.*, 1982).

Factors influencing body composition

Nutritional factors

In his literature review, Forbes (1962) described the effect of diet on body composition. Metabolic balance studies in normal infants show that the retention of nitrogen and minerals varies directly with intake: the greater the intake, the greater the concentration per unit body weight. The logical conclusion is that protein and minerals can be stored, and the concept of 'supermineralisation' has developed. On the other hand, the tissue composition of growing animals eating diets of varying protein, fat and hash content generally remains fairly constant (except for calcium) when the values are expressed per unit of fat-free tissue (Wallace *et al.*, 1958). As the authors put it, 'the composition of the body achieves an independence from the environment'. This is not surprising given the wide variation in the human diet, ranging from a high of 74% carbohydrate in India (FAO, 1991) to a low of 8% carbohydrate in Eskimos (Krogh & Krogh, 1913). The influence of nutrition on growth can be studied by comparing feeding practice and growth pattern according to energy intake and diet composition.

Energy intake

Energy intake has a major influence on growth. Well-fed animals are heavier, longer and fatter at weaning, and these differences persist even after *ad libitum* feeding (Forbes, 1962). In humans, as in animals, early feeding experience has a profound influence on later body size and composition. Garn & Clark (1975) demonstrated that obese children tend to be taller, have larger skeletal and muscular masses and have greater mineral masses than non-obese children. Most studies report similar findings (Forbes, 1964; Malina *et al.*, 1989). However, Vignolo *et al.* (1988) found that advanced stature in childhood is not maintained and may even be decreased in adults. Rusconi (1988) showed that if overweight starts in mid or late puberty, when height velocity is maximal, it can compromise final adult height.

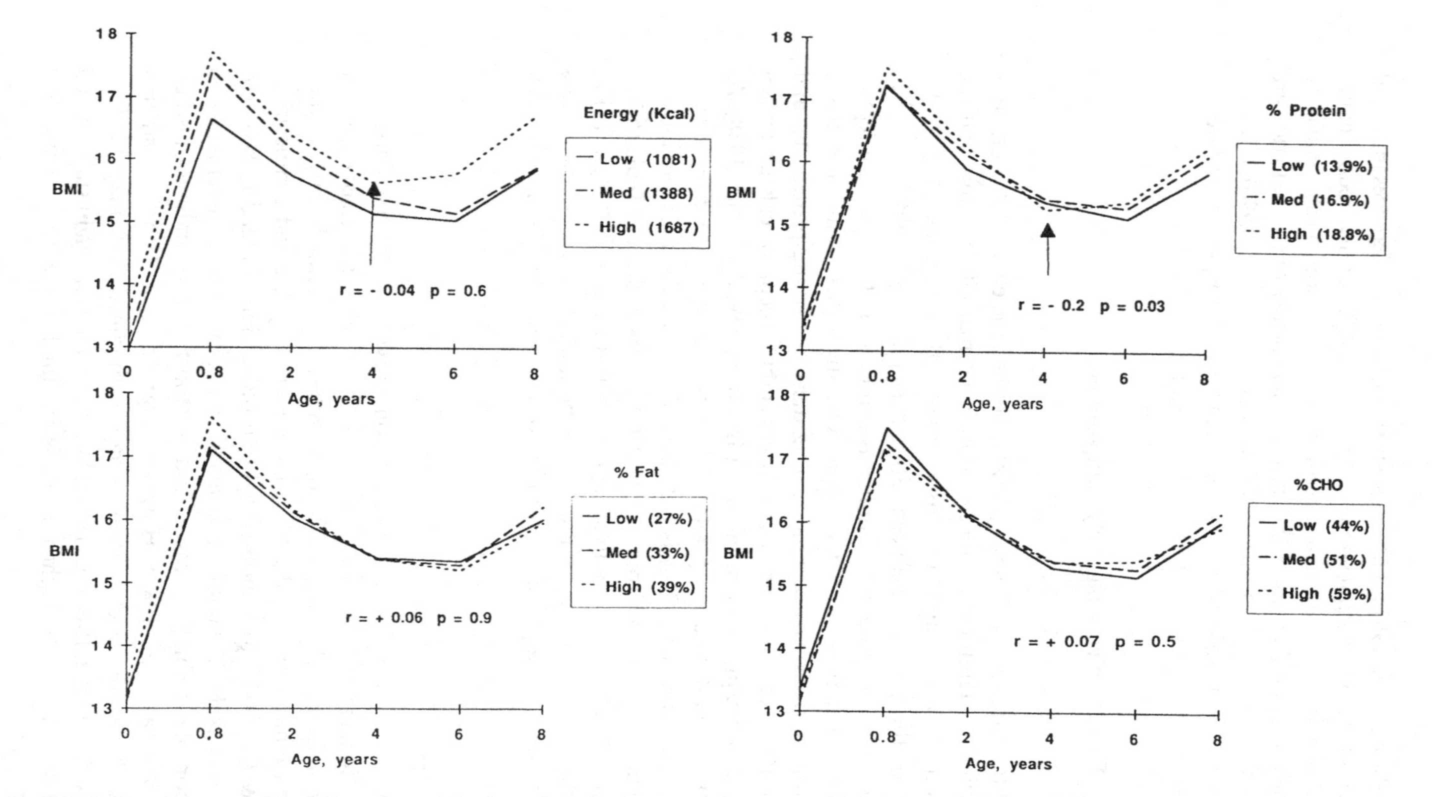

Fig. 8.7. BMI patterns according to dietary energy intake and nutrient content at 2 years ($n = 126$) and correlation coefficients between intake at 2 years and age at adiposity rebound. The arrows indicate a negative correlation, i.e. the higher the inake at 2 years, the earlier the adiposity rebound (significant for protein only) and the higher the subsequent BMI level.

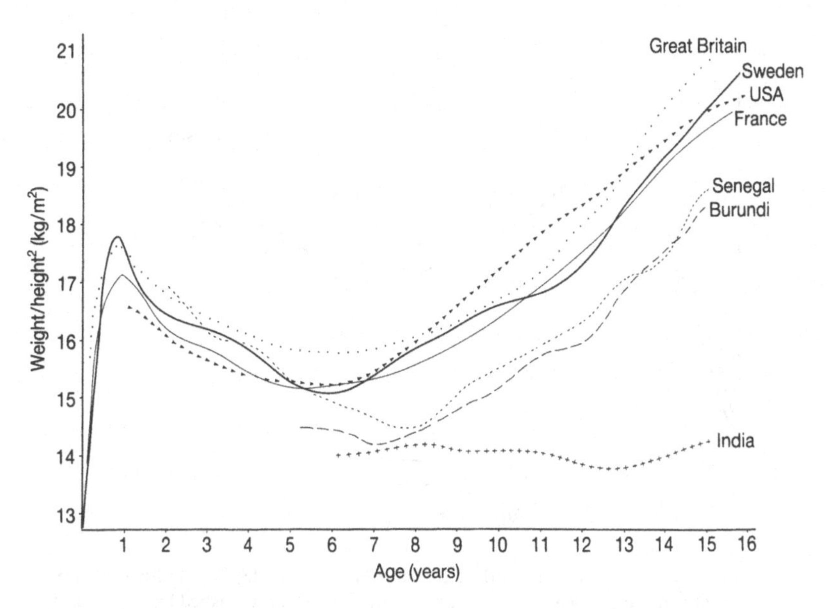

Fig. 8.8. BMI development in girls in various countries. After Rolland-Cachera (1993).

In a French longitudinal study of nutrition and growth started in 1985, we compared BMI patterns from birth to 8 years in three groups of children according to their energy intake at 2 years (Fig. 8.7). The sample and methods have already been described (Deheeger *et al.*, 1991, 1994). As expected, a high energy intake at 2 years was associated with a high BMI at 8 years, but this was apparent only after the age at adiposity rebound. Cross-sectional correlations between BMI and energy were insignificant at these two ages.

The BMI curves of countries with different nutritional environments look very different. Judged by the BMI distribution, there is a higher prevalence of obesity in the USA than in France, but only after the age of 6 years (Rolland-Cachera, 1993). Comparing mean BMI curves of USA, France and Senegal (Fig. 8.8), unexpected differences are seen during infancy: the lowest values are in the USA and the highest in Senegal. After 6 years the differences are less surprising: Senegal has the lowest values and the latest adiposity rebound (8 years compared with 6.2 in France and 5.3 in the USA). The differences before and after the rebound can be explained as a shift to the left of the curve in populations where high intake and low activity are common, and a shift to the right in populations where overeating is rare. This inverse fatness status between early and late growth was also seen when relating BMI and SF development to the age at adiposity rebound (Rolland-Cachera *et al.*, 1987): the earlier the rebound,

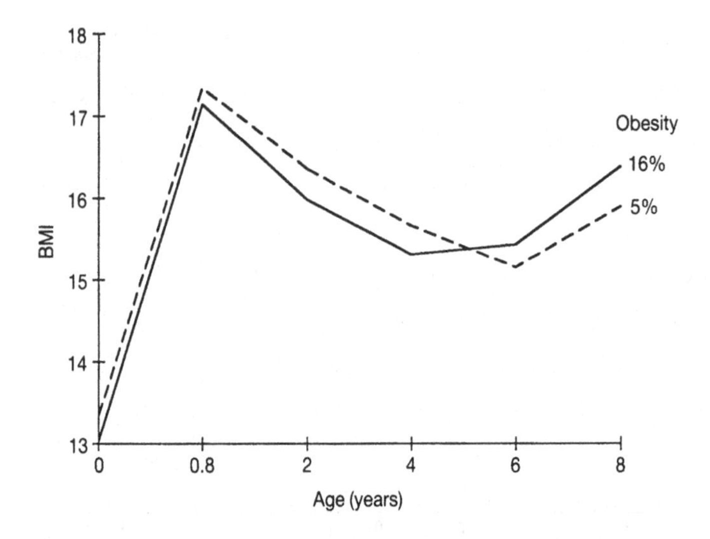

Fig. 8.9. BMI pattern according to duration of breast feeding and prevalence of obesity at the age of 8 years. Continuous line, <1 month; dashed line, >1 month.

the higher the fatness in adulthood, but the lower the fatness at the age of 1 year.

Composition of the diet

Breast feeding. Breast feeding has always had its advocates, yet it is still sometimes criticised (Forsyth, 1992). Its benefit is often assessed in terms of growth in length, length velocity or weight-for-length. It is known that healthy breast-fed infants grow less rapidly after the first 2–3 months of life than 1960s standards (which are based mainly on formula-fed infants) recommend (Dewey *et al.*, 1991; Salmenpera *et al.*, 1985). Also, stunting rather than wasting is reported in breast-fed infants (Grummer-Strawn, 1993). Using the French longitudinal study of nutrition and growth, we compared BMI patterns up to 8 years in groups of children breast-fed for less and more than 1 month, 1993). At 8 years, the proportion of obese children was smaller in the group breast-fed for longer (5% versus 16%) and the mean skinfold (SS + TRI) was lower, in spite of identical energy intakes at 10 months, 2, 4, 6 and 8 years in the two groups. The breast-fed children grew more slowly until 4 years of age and they were somewhat fatter. Comparing the BMI curves for the two groups (Fig. 8.9), the main difference is an earlier adiposity rebound in the children breast-fed for less time. An early rebound is associated with later obesity, so children with an

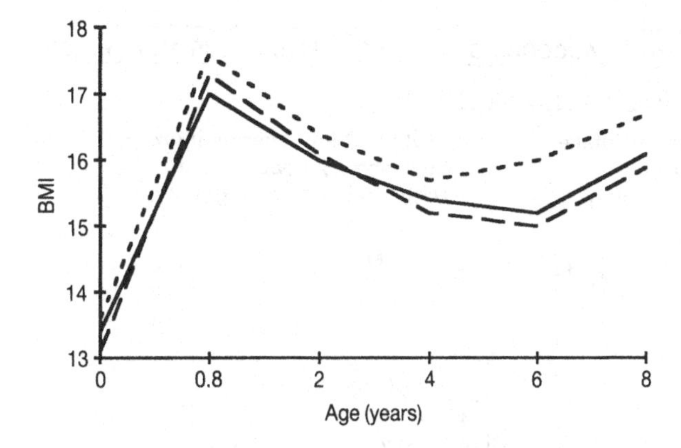

Fig. 8.10. BMI pattern according to height velocity between 2 and 4 years. Continuous line, low; long dashed line, medium; short dashed line, high height velocity.

early rapid growth phase stop growing sooner (Minuto *et al.*, 1988) and end up shorter than slow maturers (Hägg & Taranger, 1991; Cianfarani *et al.*, 1991; Prokopec, 1982). Looking at the BMI pattern in groups of children with high, medium and low height velocity before adiposity rebound (Fig. 8.10), the high-velocity group has the earliest adiposity rebound and the highest BMI at 8 years ($p<0.01$).

When assessed by short-term growth, the benefit of breast feeding is poor, but when assessed on the basis of its long-term effect, the benefits are obvious.

Influence of nutrients. Using the French longitudinal study of nutrition and growth, we have compared the BMI pattern of children according to the composition of their diet at 2 years: high, medium and low protein, fat and carbohydrate (Fig. 8.7). At no age was there a difference in energy intake between the three groups. Only protein was negatively correlated with adiposity rebound, i.e. the higher the protein content, the earlier the rebound. The mean protein content was 16% and the animal/vegetable protein ratio was approximately 3 (Deheeger *et al.*, 1991). The dietary allowance is 12% for protein and 1 for the animal/vegetable protein ratio, so that protein intake in the 'low protein' group corresponds to normal intake, while it is excessive in the other groups.

In affluent countries, a positive relationship between body fatness and energy intake can be hard to find. Conversely, high protein intake is usually recorded in the obese (Rolland-Cachera & Bellisle, 1986; Valoski & Epstein 1990; Frank *et al.*, 1978), but never discussed as a causal factor. Rapid height velocity is associated with high protein intake (Rolland-Cachera

I - HORMONAL CHANGES ACCORDING TO VARIOUS SITUATIONS AND CYCLES

PROMOTORS (some examples)

Fat, CHO, low energy intake
Exercise, cold, dark
Low pressure and gravity

Protein, high energy intake
Low energy needs, light
High pressure and gravity

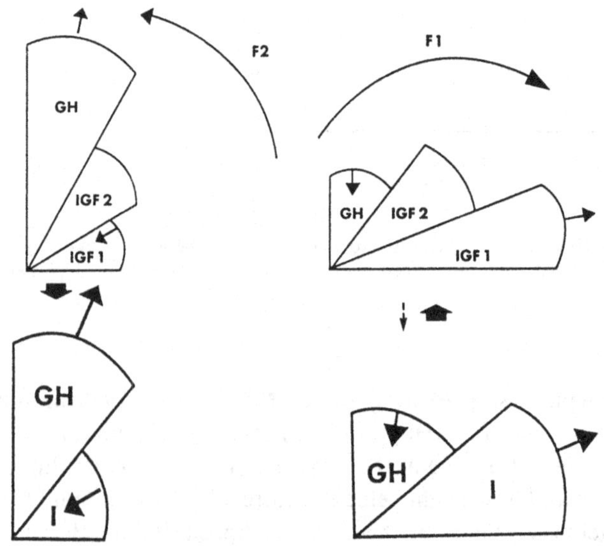

CONSEQUENCES (some examples)

↓ Protein turn over
↑ Venous pressure
↑ Tryptophan, ↑ Serotonin
Calm, maintenance period

↑ Protein turn over
↑ Blood pressure
↓ Tryptophan, ↓ Serotonin
Efficiency, stress, development

II - ABNORMAL SITUATIONS

F2 EXCESS: ↓ HPA axis →

↓ IGF1 → Poor growth and development
Stunting, Diabetes

F1 EXCESS: ↑ HPA axis →

-↑ IGFs and/or ↑ IGF2/IGF1 → Cancer
-↑ IGFs → ↓ GH (1st phase)
↓ GH → ↓ IGFs (2nd phase)
→ Obesity, Diabetes

Fig. 8.11. The concept of a balance between two opposite situations controlled by various factors: maturation (F2) and multiplication (F1). Contrasted and equilibrated F1/F2 phases may promote growth hormone (GH) secretion and good health. I, insulin; IGF1, IGF2, insulin-like growth factors; HPA, hypothalomo-pituitary-adrenal axis; CHO, carbohydrate.

et al., 1991*b*), and also with early adiposity rebound and later fatness (Fig. 8.10). The influence of protein intake has been investigated by comparing breast-fed children and infants during weaning who were fed isocaloric formulae with low and high protein content (1.3 and 1.8 g protein/100 ml) (Axelsson *et al.*, 1988). Infants in the low protein group grew more like breast-fed infants. The authors suggest that cow's-milk-based formulae should take this finding into account. The use of low-fat milk or a low carbohydrate diet also increases the protein content, so that adult weight control strategies like these should not be used early in life.

Nutrition and hormonal status

Malnutrition, including a low protein-energy diet, leads to poor growth and is associated with an IGF1 decrease, a GH increase (Maes *et al.*, 1991) and an insulin decrease (Caufriez *et al.*, 1984). The process slows down growth by stunting, limiting other health damage. The inverse process is seen in the first stage of obesity in young children: somatomedin C levels are elevated, GH is decreased and height velocity is high (Van Vliet *et al.*, 1986; Rosskamp *et al.*, 1987; Loche *et al.*, 1989). At the same time insulin is increased (Jeanrenaud *et al.*, 1992). The role of the GH/insulin ratio in controlling body fatness was first proposed by Woods *et al.* (1974).

Much evidence suggests that these hormonal changes can be predicted from BMI patterns: in breast-fed versus non-breast-fed children, or those eating a low versus high protein diet, or those living in countries with small versus large food supplies, the two groups display slow and fast growth respectively. Slow growth and late maturation associated with low levels of IGFs leads to a relatively fat infant. Waterlow (1974) described the paradoxical well-nourished clinical appearance of young children who are stunted but not wasted. Growth faltering in late maturers is probably explained by low levels of IGFs. Conversely, fast growth and early maturation increases IGFs (particularly IGF2 in early life), slimming the child but shortening the maturation phase, as shown by early adiposity rebound and advanced bone maturation (Rolland-Cachera *et al.*, 1984). The multiplication phase then rapidly increases fatness. An accelerated but shortened maturation phase in the obese leads to a paradoxical body immaturity (increased fatness, android distribution, low body density and increased extra/intracellular water) (Frisancho & Flegel, 1982; Forbes, 1987).

The protein content of the diet may have a larger effect on serum IGF1 level (Clemmons & Underwood, 1991) than does carbohydrate or fat (Snyder *et al.*, 1989). The nutrients seem to form a hierarchy in promoting IGF1, in the increasing order fat, carbohydrate and protein, consistent with their ability to increase thermogenesis (Wahren, 1992) and satiation (Blundel & Burley, 1990). The specific dynamic action (SDA) values for the different nutrients are 0–3% for fat, 5–10% for carbohydrate and 20–30 % for protein (Flatt, 1978). High protein intake increases protein loss and turnover (Price *et al.*, 1994) and may increase IGFs/insulin and IGF1/IGF2 balance promoting lipolysis and cell multiplication. Conversely, fat promotes antilipolysis but not cell multiplication. This is consistent with the observation of increased cell size, but not cell number, in 2-week-old mice suckled by mothers fed a high fat diet (Lemmonier, 1972). Increased cell number appears later, probably induced by caloric excess.

The difficulty of showing a relationship between adiposity and energy intake could be explained by these hormonal changes: because of the competition between hormones, ratios rather than absolute levels may regulate metabolism and growth. The capacity of foods to promote IGFs and modulate hormonal balance rather than energy *per se* should be important in the constitution of fat stores. It seems that all nutrients have both beneficial and detrimental effects for weight loss: a high fat diet induces hypertrophy but not hyperplasia, while high protein causes rapid weight loss due to increased thermogenesis but could be responsible for subsequent hyperplasia. The short- and long-term consequences of diet composition (affecting cell size and cell number respectively) are probably very different. Selecting carbohydrate seems a reasonable way of regulating the two opposing mechanisms.

The two phases observed in childhood overweight (thinness in infancy followed by an early rebound and later fatness), seen in situations such as formula rather than breast feeding, protein excess and positive energy balance in industrialised countries, parallel the two phases observed in obesity and diabetes development in animals (Pénicaud *et al.*, 1986; Debant *et al.*, 1987; Jeanrenaud *et al.*, 1992) as in humans (Le Stunff & Bougnères, 1992).

Fig. 8.11 presents a hypothetical view of hormonal status in two opposing situations. One corresponds to high energy needs (fasting state or low protein energy intake, cold, exercise, etc.), where IGF1 is low, GH is high and insulin is low. The second corresponds to low energy needs (high protein energy intake, warm, sedentary lifestyle, etc.), where IGF1 is high, GH is low and insulin is high. We propose naming the shift towards decreasing IGF1 the F2 route, and the shift towards increased IGF1 the F1 route. These opposing mechanisms (F2 for saving energy and F1 for increasing metabolism) provide independence from the environment necessary for survival. The normal physiological changes of hormonal status maintained by various cycles (hunger/satiety, seasons, etc.) become pathological when one route is extended at the expense of the other: in the exaggerated F2 route, decreasing IGFs lead to poor growth and development; the exaggerated F1 route increases IGF1 excessively and subsequently dries up GH, which in turn decreases both IGF1 and IGF2, as IGF2 becomes GH dependent when GH falls below normal (Zapf *et al*, 1981). Low IGF1 in both situations (F1 and F2 excess) could explain some similarities in malnutrition and obesity, such as body immaturity, the exaggerated Randle cycle (Randle *et al.*, 1963) and diabetes (Garg *et al.*, 1989; Marin *et al.*, 1993).

Food preferences and behaviour

Food intake is the outcome of specific needs as mediated through appetite. Needs change with gender, age, season and disease in particular situations. During infancy, the milk diet is sweet and the same every day. Later in life, when a wide variety of savoury meals is eaten, breakfast still tends to be sweet and the same every day. This similarity between early life and early daytime could be due to hormonal status, i.e. a low IGF1 in infancy (Rosenfeld *et al.*, 1986) and on waking (Minuto *et al.*, 1981).

A study of sensory evaluation of sugar and fat in dairy products (Monneuse *et al.*, 1991) showed that preference for very sweet foods decreases with age from infancy to adulthood. Pleasantness ratings after meals are lower in young girls than in boys. Similar age and sex differences are observed in epidemiological studies (Deheeger *et al.*, 1994): boys consume more sucrose than girls, and the intake falls with age. The 'sweet tooth' in children as compared with adults, and in boys as compared with girls, corresponds to a lower IGF1 status.

The physiological control and regulation of food intake have been reviewed by Le Magnen (1971) and Rolls (1993). Food intake is controlled by dual 'centres' (feeding and satiety) in the ventromedial hypothalamus. Sensory-specific modulation of the responsiveness of lateral hypothalamic neurons and of appetite has been demonstrated and studies have shown that the pleasantness of taste of a food eaten to satiety decreases more than that for foods that have not been eaten. This site specificity suggests a physiological memorisation of food taste during life, which is particularly important during growth. Introducing foods too soon, or conversely limiting, food exposure could impair taste learning (Birch *et al.*, 1989).

Environmental factors also affect appetite. Life-threatening situations such as food deprivation or cold switch food preferences from carbohydrate to fat or from protein to carbohydrate and fat (Sclafani & Ackroff, 1993; Leshner *et al.*, 1971; Van Staveren *et al.*, 1986). This hierarchy of preferences is the opposite of the hierarchy for promoting IGF1. Indeed, food deprivation (Clemmons & Underwood, 1991) and cold (Ma *et al.*, 1992) both decrease IGF1. Exercise is also accompanied by a reduction in plasma IGF1 (Smith *et al.*, 1987). This may explain why appetite is not increased soon after exercise (Widdowson *et al.*, 1954) and why exercise does not always lead to weight loss (Arroll & Beaglehole, 1993). In obesity (characterised by body composition immaturity and low IGF1), as in situations of high energy need, preference for fat is increased (Drewnovski *et al.*, 1992).

Food preference is also associated with sex hormones and body composition. Carbohydrate consumption increases before menstruation compared with the preovulatory period (Pliner & Fleming, 1983) when the amount of body water tends to increase due to increased progesterone (Deurenberg *et al.*, 1988). Oestrogen treatment in female rats reduces total food and carbohydrate intake but not protein intake (Wurtman & Baum, 1980). This behaviour is the opposite of high-demand situations where GH increases and IGF1 decreases, slowing down metabolism. Indeed, oestrogens increase GH (Ho & Weissberger, 1990) and probably mimic a situation of 'low need'. A similar situation probably occurs in smokers, as they have a decreased appetite for carbohydrate (Grunberg & Morse, 1984).

Palatability of food alters hormonal status. The cephalic phase of insulin secretion that occurs before ingestion (Powley, 1977) is increased by the hedonistic value of food (Louis-Sylvestre & Le Magnen, 1980), and is seen even with no energy content (Powley & Berthoud, 1985). Increased insulin before a meal occurs when GH decreases (Costin *et al.*, 1989). This is followed by an increase of IGFs (Fig. 8.11) depending on the energy and nutrient content of the food (Clemmons & Underwood, 1991) and probably on its palatability as well. This hormonal balance could explain the increased post-prandial thermogenesis following highly palatable meals (Leblanc & Brondel, 1985). A high palatability/low density food will increase insulin without increasing IGFs, limiting the advantage of 'light' food utilisation discussed by Bellisle & Perez (1994). The capacity of foods to increase IGFs and then to reduce the antilipolytic action of insulin may be an explanation of the 'French paradox' (Tunstall-Pedoe,1988): the low prevalence in France of heart disease in spite of the rich traditional cuisine. The high proportion of energy consumed later in the day by the obese (Bellisle *et al.*, 1988), at a time when IGFs hardly increase, may contribute to decreased lypolysis. Less lipolysis as opposed to more lipogenesis was proposed as the cause of obesity by Le Magnen (1985). Finally, the genetic ability to produce IGFs may also explain the differences in fatness response to overfeeding reported by Bouchard *et al.* (1990).

Another consequence of food choice is the synthesis of serotonin in the brain, which depends on the availability of its precursor amino acid tryptophan (Wurtman *et al.*, 1981). Transport of this amino acid across the blood–brain barrier competes with transport of large neutral amino acids (LNAA), so that the brain tryptophan concentration is determined by the plasma ratio of tryptophan to other LNAA. Human milk contains a higher concentration of tryptophan than cow's milk. Carbohydrate consumption (Fernstrom & Wurtman, 1971), reduced food intake (Bender, 1978; Weiss *et al.*, 1991) and exercise (Blomstrand *et al.*, 1988) all increase tryptophan and promote brain serotonin synthesis. A defect of serotonin is responsible for mood disorders in humans (Kolata, 1982). In addition, a diet poor in tryptophan increases the pain threshold (Seltzer *et al.*, 1982), and subjects consuming a high carbohydrate diet are calmer but are less efficient in performance tests (Heraief *et al.*, 1983).

Some pathologies (anorexia nervosa, carbohydrate craving) can be associated with a need for serotonin. Obesity is often associated with greater proneness to stress (Jeanrenaud *et al.*, 1992), and one biological response to stress is activation of the hypothalamo-pituitary-adrenal axis. Animals receiving repeated tail-pinch stress have a higher food intake (Antelman *et al.*, 1976), a higher locomotor response to amphetamine, and a higher intake of this drug during self administration (Piazza *et al.*, 1990). Predisposition to drug-seeking could be explained by an already existing withdrawal state created by excessive F1 versus F2. Appropriate infant feeding, and high carbohydrate consumption resulting from an active lifestyle, could promote serotonin synthesis and be factors protecting against drug seeking (see Fig. 8.11).

Influence of non-nutritional factors

Other stimuli

Various other factors may influence hormonal status and body composition. *Pressure and gravity.* Both in space and in experimental head-down tilt, microgravity or negative pressure affect biological characteristics and hormonal status, causing increased central venous pressure and decreased total peripheral resistance, calcium loss, reduction of catecholamine and an increase in cortisol (Leach & Rambaut, 1977; London *et al.*, 1983). For hormones such as thyroxine and cortisol, there appears to be a sudden transition at birth from fetal to postnatal responses (Underwood & D'Ercole, 1984) that could partly be explained by changes of gravity. Indeed, decreasing IGFs (particularly IGF1) during sleep (Minuto *et al.*, 1981) and when the fetus turns upside down suggest a role of gravity. Micro gravity increases protein catabolism and decreases protein synthesis (Goldspink *et al.*, 1986) suggesting decreased IGF1. The influence of gravity on hormonal status may also differ between supine and prone positions. This could account for the higher risk of sudden infant death syndrome in the prone position (Hunt & Schannon, 1992). Rocking has a beneficial effect, reducing the frequency of apnoeic attacks as well as bradycardia (Tuck *et al.*, 1982). Finally, low pressure could account for the smaller stature of people living in high altitude (Eveleth & Tanner, 1990). Standing position and the vertical growth of plants may correspond respectively to the maximal release of GH (and IGF1) and auxin.

Activity. Activity reduces blood pressure, decreases noradrenaline concentration and reduces peripheral vascular resistance (Arroll & Beaglehole, 1993). It also decreases insulin and increases high density lipoproteins (Lakka & Salonen, 1992) and serum potassium (Bohmer, 1986). It reduces abdominal as well as femoral adipose tissue lipoprotein lipase activity, consistent with decreased IGF1 with exercise (Smith *et al.*, 1987).

Cold. Like exercise, cold decreases IGF1 (Ma *et al.*, 1992). It improves biological characteristics and increases the noradrenaline response (Nielsen *et al.*, 1993).

Season. Many seasonal factors influence growth; for example in temperate zones height gain tends to be greater in the spring and less in the autumn (Eveleth & Tanner, 1990). This could be explained by changes in IGF1 related to temperature, food and light, all of which stimulate the hypothalamo-pituitary-adrenal (HPA) axis. Changes in light may also account for IGF1 differences between night and day.

In summary, as for nutritional status, several factors alter biological characteristics and hormonal status in two contrasting situations (Fig. 8.11): microgravity, cold and exercise, like low protein-energy (F2), decrease IGF1, while low energy needs, high gravity, light and stress, like protein excess (F1), increase IGF1. F2 increases venous pressure, and F1 increases blood pressure; F2 decreases protein synthesis and F1 increases it. Here again, both routes are necessary. Like sleep for the whole body, F2 is a metabolic 'maintenance' (GH recharge) phase, while F1 is an 'expense' phase taking advantage of positive energy balance to grow, develop and renew. A balance between the two phases is necessary. F1 excess probably increases the active phase of life explaining the disturbance of many rhythms (eating and sleep patterns) seen in obesity (Le Magnen, 1985; Bellisle *et al.*, 1988; Locard *et al.*, 1992) and high stature (Gulliford *et al.*, 1990).

Initial hormonal changes appear to be determined in the brain, affecting growth and body composition. As Jeanrenaud *et al.* (1992) said 'the origin of the syndrome of obesity is viewed as lying somewhere within the brain in which the initial disorders (due to genetic and/or environmental factors) would initiate the peripheral changes, i.e. a series of dysregulations of the autonomic nervous system and of the hypothalamo-pituitary-adrenal axis, with their pathological endocrine and metabolic consequences'. Various aspects of lifestyle and diet composition increase hormonal secretions in the first stage of obesity. This, together with an unbalanced hormonal secretion induced from the central nervous system, could be the aetiological basis of obesity. Promotion of one route rather than the other could be echoed at various levels: neurological (sympathetic versus parasympathetic), hormonal (adrenaline versus noradrenaline, IGF1/IGF2 or protein kinase/phosphatase), biological (LDL/HDL), clinical (venous/arterial pressure) or body composition (extra/intracellular water, calcium/potassium).

Not only does the environment contribute to this imbalance, but genetic factors are also probably important, as suggested by studies on familial resemblance (Stunkard *et al.*, 1990). Dietary factors such as energy and nutrient balance can either correct or aggravate an initial imbalance.

The F1/F2 classification of hormonal status matches dietary classifications seen in some traditional cultures. Dietary and therapeutic principles adopted in many populations are based on the opposing effects of foods (Centlivres, 1985). In Afghanistan (Oberlin-Faure, 1978; Centlivres, 1985) food items designated as 'hot' or 'cold' appear to correspond to the classification of nutrients according to their thermogenic effects and potency to promote F2 or F1 status. For example, almonds versus citrus, butter versus milk, or the yolk versus the white of the egg are respectively 'hot' and 'cold' when compared between items, but not on their own. This classification may involve nutrients such as fat versus protein, or vitamins such as E and A versus C, or the mineral content. It is used to achieve dietary equilibrium. Also individuals, males versus females and children versus adults are respectively designated as 'hot' and 'cold'. This classification corresponds to their low and high IGF1 status. The F2 and F1 classification also fits with the concept of duality between 'yin' and 'yang' in Ling-Shu (Ming Wong, 1987), which includes ideas of both equilibrium and thermodynamics.

GH decreases with age in adults due to a decrease in pulse frequency and, in the obese, amplitude is also reduced (Iranmanesh *et al.*, 1991). For GH secretion, as for nerve conduction (Prigogine, 1961; Changeux, 1983), the laws of thermodynamic are probably involved. An adequate dietary energy intake, varied living conditions (changes in gravity, temperature, activity and satiety), increasing F1 and F2 amplitude and an equilibrated F1/F2 balance, are probably necessary conditions to maintain both the pulse amplitude and level of GH. A defect in one or more of these conditions could be the cause of 'syndrome X' (Reaven, 1988), which is responsible for a variety of metabolic disorders.

Trends of body composition: fatness, stature, shape

Over the past 100 years children have been getting larger and reaching maturity earlier (Eveleth & Tanner, 1990). In industrialised countries, such as France and the USA, weight tends to rise faster than height, increasing the BMI (Rolland-Cachera *et al.*, 1992) and skinfold thickness (Gormaker *et al.*, 1987), particularly the highest centiles. A similar pattern has been seen in other countries (Sorensen & Price, 1990).

The trend to increasing stature is almost entirely due to an increase in leg length (Eveleth & Tanner, 1990). In two French longitudinal surveys carried out 30 years apart (Sempé, 1979; Deheeger *et al.*, 1994) there is little difference in the BMI distributions before 6 years, except for a greater prevalence of early adiposity rebound in the more recent study. By 8 years there are 3 times more obese in the later study. Stature is 2.5 cm greater at 6 years, yet sitting height at the same age is 2 cm less. This increased android pattern of statural body proportions (sitting height/stature) is seen at all ages. In addition, there is an increased android pattern of fat distribution at all ages in both sexes. This is consistent with the positive association found between advanced maturation and centripetal distribution of subcutaneous fat (Frisancho & Flegel, 1982). Compared with their birth characteristics, children are less mature for BFD and more mature for statural body proportion, which may correspond to a regression concerning GH and a progression concerning IGF1. As the different tissues (nervous system, muscle, skeleton, upper versus lower segment) grow at different rates, and achieve final growth at different ages, imbalance in the maturation and multiplication phases will affect them differently. This could create a lag between the maturational stage of the different tissues affecting final body composition.

Increased stature is reported in most cases of childhood obesity. However, the secular trend of increased stature is not always associated with obesity in individuals. It may be another response to overnutrition, mainly affecting bone tissue receptors. IGF values in constitutionally tall subjects can reach levels found in patients with active acromegaly, although no clinical or radiological evidence exists for such a diagnosis (Gourmelen *et al.*, 1984). This raises the question of whether the secular trend of increased stature is healthy, and whether it corresponds to the full expression of genetic potential. If one aspect of hormonal status (F1) develops to the detriment of another (F2), as suggested by increased leg length and decreased sitting height in children, increased stature may not be an improvement. The inverse association between growth rate and life expectancy (Micozzi, 1993) and positive association between stature and cancer (Albanes *et al.*, 1988) confirms this impression. In addition, in many countries there is a trend to increasing obesity, and obesity is associated with GH deficiency. As GH has a possible role in determining immune maturation (Bozzola *et al.*, 1989), GH deficiency may affect health in this way in addition to its other metabolic effects.

Conclusion

Although they are not precise tools for assessing body composition, anthropometric measurements are useful indicators because of their relationships with risk factors. In addition, anthropometric changes reflect changes in body composition and hormonal status. Comparisons of growth patterns suggest hypotheses about the factors that affect growth and body composition.

The birth characteristics of fatness, android body fat distribution and gynoid statural body proportion soon disappear. Adult status seems to be related to the duration of the maturation phase that occurs early in life, and that affects the various tissues differently. The duration of this phase is environmentally and nutritionally influenced, and probably accounts for the weakness of infant–adult measurement correlations. Changes during growth take place in two main periods: before and after the age of 6 years. The former is the period of chemical maturation of fat-free mass (the maturation phase or F2, associated with low IGF1), and the latter is the period of rapid cell multiplication (the multiplication phase or F1, associated with high IGF1). The two phases seem to be repeated in various cycles in life (seasonal, circadian, ovarian, etc.). F1 and F2 correspond to hormonal status, promoting either lipolysis or lipogenesis according to the context. Both situations constitute an adaptive system providing relative independence of the organism from the environment: reduced metabolism when nutrient needs are high or intakes are low, and increased metabolism when nutrient intake is high. Both routes – F2 (for maintenance and recovery) and F1 (for development and renewal) – are necessary. An imbalance arises if one develops at the expense of the other. High protein-energy intake and other modern lifestyle factors appear to promote the F1 route, which first accelerates and later shortens the maturation–maintenance phase, resulting in a paradoxical body immaturity. This could be responsible for the development of obesity.

The successive phases of growth correspond to successive phases of feeding behaviour, and the sites corresponding to different foods in hypothalamic neurons suggest the need for foods to be learned progressively. Similarly, there are site-specific functions in the brain for sight and information processing (Sperry, 1981). Learning to read, to eat, and processes such as immune function (Sterkers *et al.*, 1993) all involve memory, and so benefit from progression rather than too early diversification of stimuli. Various nutritional and psychological disorders and immunological defects could be related to accelerated growth.

In conclusion, the coincidence in the timing of different phases of growth (body composition, hormonal status and food preferences), and their changes according to various influences, suggests a relationship between hormonal status and body composition. Although supported by the scientific literature, this hypothesis needs to be confirmed. However, evidence suggests that by increasing energy expenditure, thus altering food preferences, it is possible to obtain an adequate energy and nutrient intake and stimulate GH release. Growth should be neither too slow, to take advantage of the development period, nor too rapid, to respect the successive phases of development. These simple rules should improve child growth and adult body composition. Nevertheless, even common sense advice needs persuasive arguments to make it convincing.

'To everything there is a season, and a time to every purpose under the heaven'
(ECCLESIASTES 3:1)

Acknowledgements

The author gratefully acknowledges the help of M. Sempé, M. Deheeger and F. Bellisle in completing the scientific information, and T.J. Cole for improving the manuscript.

References

Abraham, S., Collins, C., & Nordsieck, M. (1970). Relationship of weight status to morbidity in adults. *Public Health Reports*, **86**, 273–84.

Albanes, D., Jones, D.Y., Schatzkin, A., Micozzi, M.S. & Taylor, P.R. (1988). Adult stature and risk of cancer. *Cancer Research*, **48**, 1658–62.

Antelman, S.M., Rowland, N.E. & Fisher, A.E. (1976). Stimulation bound ingestive behavior: a view from the tail. *Physiology and Behaviour*, **17**, 743–8.

Arroll, B. & Beaglehole, R. (1993). Exercise for hypertension. *Lancet*, **341**, 1248–9.

Axelsson, I.E., Jakobsson, I. & Räihä, N.C.R. (1988). Formula with reduced protein content: effects on growth and protein metabolism during weaning. *Pediatric Research*, **24**, 297–301.

Bala, R.M., Lopatka, J., Leung, A., McCoy, E. & McArthur, R.G. (1981). Serum immunoreactive somatomedin levels in normal adults, pregnant women at term, children at various ages and children with constitutionally delayed growth. *Journal of Clinical Endocrinology and Metabolism*, **52**, 508–12.

Bellisle, F. & Perez, C. (1994). Low-energy substitutes for sugars and fats in human diet. Impact on nutritional regulation. *Neurosciences and Biobehavioral Reviews*, **18**, 197–205.

Bellisle, F., Rolland-Cachera, M.F., Deheeger M. & Guilloud-Bataille, M. (1988).

Obesity and food intake in children: evidence for a role of metabolic and/or behavioral daily rhythms. *Appetite*, **11**, 11–18.
Bender, D.A. (1978). Regulation of 5-hydroxytryptamine synthesis. *Proceedings of the Nutrition Society*, **37**, 159–65.
Binoux, M., Hossenlopp, C., Lassare, C. & Hardouin, N. (1981). Production of insulin-like growth factors and their carrier by rats pituitary gland and brain explants in culture. *FEBS Letters*, **124**, 178–84.
Birch, L.L., McPhee, L., Sullivan, S. & Johnson, S. (1989). Conditioned meal initiation in young children. *Appetite*, **13**, 105–13.
Blomstrand, E., Celsing, F. & Newsholme, E.A. (1988). Changes in plasma concentrations of aromatic and branched-chain amino acids during sustained exercise in man and their possible role in fatigue. *Acta Physiologica Scandinavica*, **133**, 115–21.
Blundell, J.E. & Burley, V.J. (1990). Evaluation of the satiating power of dietary fat in man. In *Progress in Obesity Research*, ed. Y. Oomura, pp. 453–7. London: John Libbey.
Bohmer, D. (1986). Loss of electrolytes by sweat in sports. In *Sport, Health and Nutrition*, ed. F.I. Katch, pp. 67–74. Champaign, Illinois: Human Kinetics.
Bouchard, C., Tremblay, A., Després, J.P. *et al.* (1990). The response to long-term overfeeding in identical twins. *New England Journal of Medicine*, **322**, 1477–82.
Bozzola, M., Cisternino, M., Valtorta, A., *et al.* (1989). Effect of biosynthetic methionyl growth hormone (GH) therapy on the immune function in GH-deficient children. *Hormone Research*, **31**, 153–6.
Brauner, R. & de Zegher, F. (1993). Croissance et maturation foetale. *Médecine et Sciences*, **9**, 271–6.
Brook, C.G.D. (1973). Effect of human growth hormone treatment on adipose tissue in children. *Archives of Disease in Childhood*, **48**, 725–8.
Caufriez, A., Golstein, J., Lebrun, P., Herchuelz, A., Furlanetto, R. & Copinschi, G. (1984). Relations between immunoreactive somatomedin C, insulin and T3 patterns during fasting in obese subjects. *Clinical Endocrinology*, **20**, 65–70.
Centlivres, P. (1985). Hippocrate dans la cuisine: le chaud et le froid en Afghanistan du Nord. In *Identité alimentaire et altérité culturelle*, p. 35–58. Actes du colloque de Neuchâtel 12–13 Nov 1984. Neuchâtel: Institut d'Ethnologie de Neuchâtel (Suisse).
Changeux, J.P. (1983). *L'homme Neuronal.* Paris: Fayard.
Cianfarani, S., Vaccaro, F., Spadoni, G.L. *et al.* (1991). Abnormal growth velocity does not impair the attainment of target height in 'small-delay' children. *Acta Medica Auxologica*, **23**, 245–51.
Clemmons, D.R. & Underwood, L.E. (1991). Nutritional regulation of IGF-1 and IGF binding proteins. *Annual Reviews of Nutrition*, **11**, 393–412.
Cole, T.J. (1986). Weight/height^p compared to weight/height^2 for assessing adiposity in childhood: influence of age and bone age on p during puberty. *Annals of Human Biology*, **13**, 433–51.
Cornblatt, M., Parker, M.L., Reisner S.H., Forbes, A.E. & Daughaday, H. (1965). Secretion and metabolism of growth hormone in premature and full-term infants. *Journal of Clinical Endocrinology*, **25**, 209–18.
Costin, G., Kaufman, F.R. & Brasel J.A. (1989). Growth hormone secretory

dynamics in subjects with normal stature. *Journal of Pediatrics*, **115**, 537–44.
Cronk, C.E., Roche, A.F., Kent, R. *et al.* (1982). Longitudinal trends and continuity in wt/stature^2 from 3 months to 18 years. *Human Biology*, **54**, 729–49.
Cureton, K.J., Boileau, R.A. & Lohman, T.G. (1975). A comparison of densitometric, potassium-40 and skinfold estimates of body composition in prepubescent boys. *Human Biology*, **47**, 321–36.
Debant, A., Guerre-Millo, M., Le Marchand-Brustel, Y., Freychet, P., Lavau, M. & Van Obberghen, E. (1987). Insulin receptor kinase is hyper-responsive in adipocytes of young obese Zucker rats. *American Journal of Physiology*, **252**, E273–8.
Deheeger, M., Rolland-Cachera, M.F., Péquignot, F., Labadie, M.D. & Rossignol C. (1991). Evolution de l'alimentation des enfants de 2 ans entre 1973 et 1986. *Annals of Nutrition and Metabolism*, **35**, 132–40.
Deheeger, M., Rolland-Cachera, M.F., Labadie, M.D. & Rossignol, C. (1994). Etude longitudinale de la croissance et de l'alimentation d'enfants examinés de l'âge de 10 mois à 8 ans. *Cahiers de Nutrition et Diététique*, **29**, 1–8.
Deurenberg, P., Westrate, J.A., Paymans, I. & van der Kooy, K. (1988). Factors affecting bioelectrical impedance measurements in humans. *European Journal of Clinical Nutrition*, **42**, 1017–22.
Dewey, K.G., Heinig, M.J., Nommsen L.A. & Lönnerdal, B.O. (1991). Adequacy of energy intake among breast-fed infants in the Darling study: relationships to growth velocity, morbidity and activity levels. *Journal of Pediatrics*, **119**, 538–47.
Drewnowski, A., Kurth, C., Holden-Wiltse, J. & Saari, J. (1992). Food preferences in human obesity: carbohydrates versus fats. *Appetite*, **18**, 207–21.
Ducimetière, P., Richard, J. & Cambien, F. (1986). The pattern of subcutaneous fat distribution in middle-aged men and the risk of coronary heart disease: the Paris prospective study. *International Journal of Obesity*, **10**, 229–40.
Eveleth, P. & Tanner, J.M. (1990). *Worldwide Variation in Human Growth*, 2nd edn. Cambridge: Cambridge University Press.
FAO (1991). Food balance sheets. Basic data unit, pp. 157. Rome: Statistics Division.
Fernstrom, J.D. & Wurtman, R.J. (1971). Brain serotonine content: increase following ingestion of carbohydrate diet. *Science*, **174**, 1023–5.
Filipovsky, J., Ducimetière, P., Darné, B. & Richard, L. (1993). Abdominal body fat distribution and elevated blood pressure are associated with increased risk of death from cardiovascular diseases and cancer in middle-aged men. The results of a 15- to 20-years follow-up in the Paris prospective study 1. *International Journal of Obesity*, **17**, 197–203.
Flatt, J.P. (1978). The biochemistry of energy expenditure. In *Recent Advances in Obesity Research*, vol 2, ed. G.A. Bray, pp. 211–28. London: Newman.
Fomon, S.J., Haschke, F., Ziegler, E.E. & Nelson, S.E. (1982). Body composition of reference children from birth to age 10 years. *American Journal of Clinical Nutrition*, **35**, 1169–75.
Forbes, G.B. (1962). Methods for determining composition of the human body. *Pediatrics*, **29**, 477–94.
Forbes, G.B. (1964). Lean body mass and fat in obese children. *Pediatrics*, **34**, 308–14.
Forbes, G.B. (1972). Growth of the lean body mass in man. *Growth*, **36**, 324–38.
Forbes, G.B. (1977). Nutrition and growth. *Journal of Pediatrics*, **91**, 40–2.

Forbes, G.B. (1978). Body composition in adolescence. In *Human Growth*, vol 2, ed. F. Falkner & J.M. Tanner, pp. 119–39. London: Plenum Press.
Forbes, G.B. (1987). *Human Body Composition*. New York: Springer-Verlag.
Forsyth, J.S. (1992). Is it worthwhile breast-feeding? *European Journal of Clinical Nutrition*, **46** (Suppl 1), S19–25.
Fox, D., Peters, D., Armstrong, M., Sharpe, P. & Bell, M. (1993). Abdominal fat deposition in children. *International Journal of Obesity*, **17**, 11–16.
Frank, G.C., Berenson, G.S. & Webber, L.S. (1978). Dietary studies and the relationship of diet to cardiovascular disease risk factor variables in 10-year-old children: the Bogalusa heart study. *American Journal of Clinical Nutrition*, **31**, 328–40.
Freedman, D.S., Srinivasan, S.R., Burke, G.L. *et al.*, (1987). Relation of body fat distribution to hyperinsulinemia in children and adolescents: the Bogalusa heart study. *American Journal of Clinical Nutrition*, **46**, 403–10.
Frisancho, R. & Flegel, P.N. (1982). Advanced maturation associated with centripetal fat pattern. *Human Biology*, **54**, 717–27.
Garg, S.K., Marwaha, R.K., Ganpathy, V. *et al.*, (1989). Serum growth hormone, insulin and blood sugar responses to oral glucose in protein energy malnutrition. *Tropical and Geographical Medicine*, **41**, 9–13.
Garn, S.M. & Clark, D.C. (1975). Nutrition, growth, development and maturation. *Pediatrics*, **56**, 306–19.
Garrow, J.S. & Webster, J. (1985). Quetelet's index (W/H^2) as a measure of fatness. *International Journal of Obesity*, **9**, 147–53.
Girard, J. (1989). Hormonal regulation of fetal growth. In *Intrauterine growth retardation*, ed. J. Santerre, pp. 23–37. New York: Raven Press.
Girardet, J.P., Tounian, P., Le Bars, M.A. & Boreux, A. (1993). Obesité de l'enfant: intérêt des indicateurs cliniques d'évaluation. *Annales de Pédiatrie*, **40**, 297–303.
Goldspink, D.F., Morton, A.J., Loughna, P. & Goldspink, G. (1986). The effect of hypokinesia and hypodynamia on protein turnover and the growth of four skeletal muscles of the rat. *European Journal of Physiology*, **407**, 333–40.
Gormaker, S.L., Dietz, W.H. Jr, Sobom, A.M. & Wehler, C.A. (1987). Increasing pediatric obesity in the United States. *American Journal of Diseases in Childhood*, **141**, 535–40.
Gourmelen, M., Le Bouc, Y., Girard, F., *et al.* (1984). Serum levels of insulin-like growth factor (IGF) and IGF binding protein in constitutionally tall children and adolescents. *Journal of Clinical Endocrinology and Metabolism*, **59**, 1197–203.
Grummer-Strawn, L.M. (1993). Does prolonged breast-feeding impair child growth? A critical review. *Pediatrics*, **91**, 766–71.
Grunberg, N.E. & Morse, D.E. (1984). Cigarette smoking and food consumption in the United States. *Journal of Applied Sociology and Physiology*, **14**, 310–7.
Guillaume-Gentil, C., Rohner-Jeanrenaud, F., Abramo, F., Bestetti, G.E., Rossi, G.L. & Jeanrenaud, B. (1990). Abnormal regulation of the hypothalamo-pituitary-adrenal axis in the genetically obese fa/fa rat. *Endocrinology*, **126**, 1873–9.
Gulliford, M.C., Price, C.E., Rona, R.J. & Chinn, S. (1990). Sleep habits and height at ages 5 to 11. *Archives of Disease in Childhood*, **65**, 119–22.
Hägg, U. & Taranger, J. (1991). Height and height velocity in early, average and late maturers followed to the age of 25: a prospective longitudinal study of

Swedish urban children from birth to adulthood. *Annals of Human Biology*, **18**, 47–56.
Hall, K. & Sara, V.R. (1984). Somatomedin levels in childhood, adolescence and adult life. *Clinics in Endocrinology and Metabolism*, **13**, 91–112.
Henderson, S.A. (1968). Un examen épidémiologique de la maladie psychiatrique de l'adolescent. *Psychiatrie Enfant*, **XI**, 269–96.
Heraief, E., Burckhard, H., Mauron, C., Wurtman, J.J. & Wurtman R.J. (1983). The treatment of obesity by carbohydrate deprivation suppresses plasma tryptophan and its ratio to other large neutral amino acids. *Neural Transmission*, **57**, 187–95.
Himes, J.H. (1988). Alteration in distribution of body fat tissue in response to nutritional intervention. In *Fat Distribution During Growth and Later Outcomes*, ed. C. Bouchard & F.E. Johnston, pp 313–32. New York: Alan Liss.
Himes, J. & Bouchard, C. (1989). Validity of anthropometry in classifying youths as obese. *International Journal of Obesity*, **13**, 183–93.
Ho, K.Y. & Weissberger, A.J. (1990). Secretory patterns of growth hormone according to sex and age. *Hormone Research*, **33** (Suppl 4), 7–11.
Hunt, C. & Schannon, D. (1992). Sudden infant death syndrome and sleeping position. *Pediatrics*, **90**, 115–7.
Hutt, C. (1972). Sex differences in human development. *Human Development*, **15**, 153–70.
Iranmanesh, A., Lizarralde, G. & Veldhuis, J.D. (1991). Age and relative adiposity are specific negative determinants of the frequency and amplitude of growth hormone (GH) secretory bursts and the half-life of endogenous GH in healthy men. *Journal of Clinical Endocrinology and Metabolism*, **73**, 1081–8.
Jeanrenaud, B. (1979). Insulin and obesity. *Diabetologia*, **17**, 133–8.
Jeanrenaud, B., Rohner-Jeanrenaud, F., Cusin I. *et al.*, (1992). The importance of the brain in the aetiology of obesity and type 2 diabetes. *International Journal of Obesity*, **16**, S9–12.
Joly, R. (1978). Hippocrate Tome XIII. Du système des glandes, pp. 104–29. Paris: Les Belles Lettres.
Keys, A., Fidanza, F., Karvonen, M.J., Kimura, N. & Taylor, H.L. (1972). Indices of relative weight and obesity. *Journal of Chronical Diseases*, **25**, 329–43.
Knittle, J.L., Timmers, K., Ginsberg-Fellner, F., Brown, R.E. & Katz, D.P. (1979). The growth of adipose tissue in children and adolescents. Cross-sectional and longitudinal studies of adipose cell number and size. *Journal of Clinical Investigation*, **63**, 239–46.
Kolata, G. (1982). Food affects human behaviour. *Science*, **218**, 1209–10.
Krogh, A. & Krogh, M. (1913). A study of the diet and metabolism of Eskimos undertaken in 1908 on an expedition to Greenland. *Medd om Grønland*, **51**, 1–52.
Lakka, T.A. & Salonen, J.T. (1992) Physical activity and serum lipids: a cross-sectional population study in eastern Finnish men. *American Journal of Epidemiology*, **136**, 806–18.
Lapidus, L., Bengtsson, C., Larson, B. *et al.* (1984). Distribution of adipose tissue and risk of cardiovascular diseases and death. *British Medical Journal*, **289**, 1261–3.
Lapidus, L., Helgesson, Ö., Merck, C. & Bjorntorp, P. (1988). Adipose tissue

distribution and female carcinomas. A 12-years follow-up of participants in the population study of women in Gothenburg, Sweden. *International Journal of Obesity*, **12**, 361–8.

Lassare, C., Hardouin, S., Daffos, F., Forestier, F., Frankenne, F. & Binoux, M. (1991). Serum insulin-like growth factors and insulin-like growth factor binding proteins in the human fetus. Relationships with growth in normal subjects and in subjects with intrauterine growth retardation. *Pediatric Research*, **29**, 219–25.

Le Magnen, J. (1971). Advances in studies on the physiological control and regulation of food intake. In *Progress in Physiological Psychology*, vol 4, ed. E. Stellar & J.M. Sprague. New York: Academic Press.

Le Magnen, J. (1985). *Hunger*. Cambridge: Cambridge University Press.

Le Stunff, C. & Bougnères, P.F. (1992). Glycerol production and utilization during the early phase of human obesity. *Diabetes*, **41**, 444–50.

Leach, C.S. & Rambaut, P.C. (1977). Biochemical responses of the Skylab crewmen: an overview. In *Biochemical Results of Skylab*, ed. R.S. Johnston & L.F. Dietlein, pp. 204–16. Washington, D.C.: NASA SP-377.

Leblanc, J. & Brondel, L. (1985). Role of palatability on meal-induced thermogenesis in human subjects. *American Journal of Physiology*, **248**, E333–6.

Lemonnier, D. (1972). Age and sex and site on the cellularity of the adipose tissue in mice and rats rendered obese by a high-fat diet. *Journal of Clinical Investigation*, **51**, 2907–15.

Leshner, A., Collier, G. & Squibb, R.L. (1971). Dietary self-selection at cold temperature. *Physiology and Behaviour*, **6**, 1–3.

Lesko, S.M., Rosenberg, L. & Shapiro, S. (1993). A case–control study of baldness in relation to myocardial infarction in men. *Journal of the American Medical Association*, **269**, 998–1003.

Locard, E., Mamelle, N., Billette, A., Miginiac, M., Munoz, F. & Rey, S. (1992). Risk factors of obesity in a five-year-old population. Parental versus environmental factors. *International Journal of Obesity*, **16**, 721–9.

Loche, S., Pintor, C., Cappa, M., Ghigo, E., Puggioni, R., Locatelli, V. & Muller, E.E. (1989). Pyridostigmine counteracts and blunts growth hormone response to growth hormone-releasing hormone of obese children. *Acta Endocrinologica*, **120**, 624–8.

Lohman, T.G., Boileau, R.A. & Slaughter, M.H. (1984). Body composition in children and youth. In *Advances in Pediatric Sports Sciences*, ed. R.A. Boileau, pp. 29–57. Champaign, Illinois: Human Kinetics.

London, G.M., Levenson, J.A., Safar, M.E., Simon, A.C., Guerin, A.P. & Payen, D. (1983). Hemodynamic effects of head-down tilt in normal subjects and sustained hypertensive patients. *American Journal of Physiology*, **245**, 194–202.

Louis-Sylvestre, J. & Le Magnen, J. (1980). Palatability and preabsorptive insulin release. *Neuroscience and Biobehavioral Reviews*, **4**(Suppl 1), 43–6.

Ma, L., Burton K.A., Saunders, J.C. & Dauncey, M.J. (1992). Thermal and nutritional influences on tissue levels of insulin-like growth factor-1 mRNA and peptides. *Journal of Thermal Biology*, **2**, 89–95.

Maes, M., Maiter, D., Thissen, J.P., Underwood, L.E. & Ketelslegers, J.M. (1991). Contributions of growth hormone receptors and postreceptors defects to growth hormone resistance in malnutrition. *Trends of Endocrinology and Metabolism*, **2**, 92–7.

Malina, R.M., Skarabanek, M.F. & Little, B. (1989). Growth and maturity status in black and white children classified as obese by different criteria. *American Journal of Human Biology*, **1**, 193–9.

Marin, P., Kvist, H., Lindstedt, G., Sjostrom, L. & Bjorntorp, P. (1993). Low concentrations of insulin-like growth factor-I in abdominal obesity. *International Journal of Obesity*, **17**, 83–9.

Marquardt, H., Todaro, G.J., Henderson, L.E. & Oroszlan, S. (1981). Purification and primary structure of a polypeptide with multiplication-stimulating activity from rat liver cell cultures. *Journal of Biological Chemistry*, **256**, 6859–65.

Merimée, T.J., Zapf, J. & Froesch, E.R. (1982). Insulin-like growth factors in the fed and fasted states. *Journal of Clinical Endocrinology and Metabolism*, **55**, 999–1002.

Micozzi, M.S. (1993) Functional consequences from varying patterns of growth and maturation during adolescence. *Hormone Research*, **39**(Suppl 3), 49–58.

Ming Wong (1987). *Ling-Shu*. Paris: Masson.

Minuto, F., Underwood, L.E., Grimaldi, P., Furlanetto, R.W., Van Wyk, J.J. & Giordano, G. (1981). Decreased serum somatomedin C concentration during sleep: temporal relationship to the nocturnal surges of growth hormone and prolactin. *Journal of Clinical Endocrinology and Metabolism*, **52**, 399–403.

Minuto, F., Barreca, A., Del-Monte, P., Fortini, P., Resentini, M., Morabito, F. & Giordano, G. (1988). Spontaneous growth hormone and somatomedin-C/insulin-like growth factor-1 secretion in obese subjects during puberty. *Journal of Endocrinological Investigation*, **11**, 489–95.

Monneuse, M.O., Bellisle, F. & Louis-Sylvestre, J. (1991). Impact of sex and age on sensory evaluation of sugar and fat in dairy products. *Physiology and Behavior*, **50**, 1111–17.

Moulton, C. R. (1923). Age and chemical development in mammals. *Journal of Biological Chemistry*, **57**, 79–97.

Must, A., Jacques, P.F., Dallal, G.E., Bajema, C.J. & Dietz, W.H. (1992). Long-term morbidity and mortality of overweight adolescents – a follow-up of the Harvard Growth Study of 1922 to 1935. *New England Journal of Medicine*, **327**, 1350–5.

Naeye, L.R. & Roode, P. (1970). The size and number of cells in visceral organs in human obesity. *American Journal of Clinical Pathology*, **54**, 251–8.

Nichols, B.L., Hazlewood, C.F. & Barnes, D.J. (1968). Percutaneous needle biopsy of quadriceps muscle: potassium analysis in normal children. *Journal of Pediatrics*, **72**, 840–6.

Nielsen, B., Astrup, A. Samuelsen, P., Wengholt, H. & Christensen, N.J. (1993). Effect of physical training on thermogenic responses to cold and ephedrine in obesity. *International Journal of Obesity*, **17**, 383–90.

Oberlin-Faure, O. (1978). *L'alimentation traditionnelle en Afghanistan: enquête nutritionnelle et essai d'application à un projet d'éducation sanitaire*. Paris: Faculté de médecine Pitié Salpétrière.

Pénicaud, L., Rohner-Jeanrenaud, F. & Jeanrenaud, B. (1986). *In vivo* metabolic changes as studied longitudinally after ventromedial-hypothalamic lesions. *American Journal of Physiology*, **250**, E662–8.

Pertzelan, A., Keret, R., Bauman, B. *et al.*, (1986). Responsiveness of pituitary hGH to GRH 1-44 in juveniles with obesity. *Acta Endocrinologica*, **111**, 151–3.

Piazza, P.V., Demière, J.M., Le Moal, M. & Simon, H. (1990). Stress- and pharmacologically induced behavioral sensitization increases vulnerability to acquisition of amphetamine self-administration. *Brain Research*, **514**, 22–6.

Pliner, P. & Fleming, A.S. (1983). Food intake, body weight and sweetness preference over the menstrual cycle. *Physiology and Behavior*, **30**, 663–6.

Poskitt, E.M.E. & Cole, T.J. (1977). Do fat babies stay fat? *British Medical Journal*, **i**, 7–9.

Powley, T.L. (1977). The ventromedial hypothalamic syndrome, satiety and cephalic phase hypothesis. *Psychological Review*, **84**, 89–126.

Powley, T.L. & Berthoud, A.R. (1985). Diet and cephalic phase insulin response. *American Journal of Clinical Nutrition*, **42**, 991–1002.

Price, G.M., Halliday, D., Pacy, P.J., Quevedo, R. & Millward, D.J. (1994). Nitrogen homoeostasis in man: influence of protein intake on the amplitude of diurnal cycling of body nitrogen. *Clinical Science*, **86**, 91–102.

Prigogine, I. (1961). *Introduction to the Thermodynamics of Irreversible Processes*. New York: Interscience.

Prokopec, M. (1982). Early and late maturers. *Anthropologischer Közl*, **26**, 13–24.

Prokopec, M. & Bellisle, F. (1993). Adiposity in Czech children followed from 1 month of age to adulthood: analysis of individual BMI patterns. *Annals of Human Biology*, **20**, 517–25.

Ramirez, M.E. (1993). Familial aggregation of subcutaneous fat deposits and the peripheral fat deposition pattern. *International Journal of Obesity*, **17**, 63–8.

Randle, P.J., Garland, P.B., Hales, C.N. & Newsholme, E.A. (1963). The glucose fatty acids cycle. Its role in insulin sensitivity and the metabolic disturbance of diabetes mellitus. *Lancet*, i, 785–9.

Reaven, G.M. (1988). Role of insulin resistance in human disease. *Diabetes*, 37, 1595–607.

Rebuffé-Scrive, M., Enk, L., Crona, N., Lönroth, P., Abrahamsson, L., Smith, U. & Björntorp, P. (1985). Fat cell metabolism in different regions in women. Effects of menstrual cycles, pregnancy and lactation. *Journal of Clinical Investigation*, **75**, 1973–6.

Rebuffé-Scrive, M., Eldh, J., Hafström, L.O. & Björntorp P. (1986). Metabolism of mammary, abdominal and femoral adipocytes in women before and after menopause. *Metabolism*, **35**, 792–7.

Roche, A.F., Siervogel, R.M., Chumlea, W.B. & Webb, P. (1981). Grading body fatness from limited anthropometric data. *American Journal of Clinical Nutrition*, **34**, 2831–8.

Rolland-Cachera, M.F. (1993). Body composition during adolescence: methods, limitations and determinants. *Hormone Research*, **39**(Suppl 3), 25–40.

Rolland-Cachera, M.F. & Bellisle, F. (1986). No correlation between adiposity and food intake: why are working class children fatter? *American Journal of Clinical Nutrition*, **44**, 779–87.

Rolland-Cachera, M.F., Sempé, M., Guilloud-Bataille, M., Patois, E., Péquignot-Guggenbuhl, F. & Fautrad, V. (1982). Adiposity indices in children. *American Journal of Clinical Nutrition*, **36**, 178–84.

Rolland-Cachera, M.F., Deheeger, M., Bellisle, F., Sempé, M., Guilloud-Bataille, M. & Patois E. (1984). Adiposity rebound in children: a simple indicator for predicting obesity. *American Journal of Clinical Nutrition*, **39**, 129-35.

Rolland-Cachera, M.F., Deheeger, M., Avons, P., Guilloud-Bataille, M., Patois, E. & Sempé, M. (1987). Tracking adiposity patterns from 1 month to adulthood. *Annals of Human Biology*, **14**, 219–22.

Rolland-Cachera, M.F., Bellisle, F. & Sempé, M. (1989). The prediction in boys and girls of the weight/height^2 index and various skinfolds measurements in adults: a two-decade follow-up study. *International Journal of Obesity*, **13**, 305–11.

Rolland-Cachera, M.F., Bellisle, F., Péquignot, F., Deheeger, M. & Sempé, M. (1990). Influence of body fat distribution during childhood on body fat distribution during adulthood: a two-decade follow-up study. *International Journal of Obesity*, **14**, 473–81.

Rolland-Cachera, M.F., Cole, T.J., Sempé, M., Tichet, J., Rossignol, C. & Charraud, A. (1991*a*). Body mass index variations: centiles from birth to 87 years. *European Journal of Clinical Nutrition*, **45**, 13–21.

Rolland-Cachera, M.F., Deheeger, M., Péquignot, F., Labadie, M.D. & Rossignol, C. (1991*b*). Relationship between nutrition and height velocity. VIth International Congress of Auxology Abstracts, p. 49. Madrid: Fundaçion Eizaguirre.

Rolland-Cachera, M.F., Spyckerelle, Y. & Deschamps, J.P. (1992). Evolution of pediatric obesity in France. *International Journal of Obesity*, **16**(Suppl 1), 5.

Rolls, E.T. (1993). The neural control of feeding in primates. In *Neurophysiology of Ingestion*, ed. D.A. Booth, pp. 137–69. Oxford: Pergamon.

Rosenbaum, M., Gertner, J.M. & Leibel, R.L. (1989). Effect of systemic growth hormone administration on regional adipose tissue distribution and metabolism in GH-deficient children. *Journal of Clinical Endocrinology and Metabolism*, **69**, 1274–81.

Rosenfeld, R.G., Wilson, D.M., Lee, P.D.K. & Hintz, R.L. (1986). Insulin-like growth factor 1 and 2 in evaluation of growth retardation. *Journal of Pediatrics*, **109**, 428–33.

Rosskamp, R., Becker, M. & Soetadji, S. (1987). Circulating somatomedin-C levels and the effect of growth releasing factor on plasma levels of growth hormone and somatomedin-like immunoreactivity in obese children. *European Journal of Pediatrics*, **146**, 48–50.

Rusconi, R. (1988). Effects of increased body mass on growth and skeletal maturation. *Acta Medica Auxologica*, **20**, 141–52.

Salmenpera, L., Perheentura, J. & Siimes, M.A. (1985). Exclusively breast-fed healthy infants grow slower than reference infants. *Pediatric Research*, **3**, 307–12.

Sangi, H. & Mueller, W.H. (1991). Which measure of body fat distribution is best for epidemiological research among adolescents? *American Journal of Epidemiology*, **133**, 870–83.

Sara, V.R., Hall, K. & Wetterberg, L. (1981). Fetal brain growth: a proposed model for regulation by embrionic sometomedin. *Biology and Human Growth*, **7**, 241–52.

Sara, V.R., Hall, K., Enzel, K. *et al.* (1982) Somatomedin in ageing and dementia disorders of Alzheimer type. *Neurobiology of Aging*, **3**, 117–20.

Sclafani, A. & Ackroff, K. (1993). Deprivation alters rats' flavor preferences for carbohydrates and fats. *Physiology & Behavior*, **53**, 1091–9.

Seidell, J.C., Oosterle, A. & Thijssen, M.A.O. (1987). Assessment of intra-abdominal and subcutaneous abdominal fat – relation between anthropometry and computed tomography. *American Journal of Clinical Nutrition*, **45**, 7–13.

Seltzer, S., Stoch, R., Marcus, R. & Jackson, E. (1982). Alteration of human pain thresholds by nutritional manipulation and L tryptophan supplementation. *Pain*, **13**, 381–93.

Sempé, M. (1988). L'analyse de la maturation squeletique du poignet de la main et du coude. In *La Pediatrie au Quotidien*, ed. P. Doin. Paris: Les Editions INSERM.

Sempé, M., Pédron, G. & Roy-Pernot, M.P. (1979). *Auxologie, Méthode et Séquences*. Paris: Théraplix.

Siervogel, R.M., Roche, A.F., Himes, J.H., Chumlea, W.C. & McCammon, R. (1982). Subcutaneous fat distribution in males and females from 1 to 39 years of age. *American Journal of Clinical Nutrition*, **36**, 162–71.

Siervogel, R.M., Roche, A.F., Guo, S., Mukherjee, D. & Chumlea W.C. (1991). Patterns of change in weight/stature^2 from 2 to 18 years: findings from long-term serial data for children in the Fels longitudinal growth study. *International Journal of Obesity*, **15**, 479–85.

Smith, A.T., Clemmons, D.R., Underwood L.E., Ben-Ezra V. & McMurray R. (1987). The effect of exercise on plasma somatomedin-C/insulin like growth factor 1 concentration. *Metabolism*, **36**, 533–7.

Smith, U., Hammarsten, J., Björntorp, P. & Kral, J. (1979). Regional differences and effects of weight reduction on human fat cell metabolism. *European Journal of Clinical Investigation*, **9**, 327–32.

Snyder, D.K., Clemmons, D.R. & Underwood, L.E. (1989). Dietary carbohydrate content determines responsiveness to growth hormone in energy restricted humans. *Journal of Clinical Endocrinology and Metabolism*, **69**, 745–52.

Sorensen, T.I.A. & Price, R.A. (1990). Secular trends in body mass index among Danish young men. *International Journal of Obesity*, **14**, 411–19.

Sperry, R.W. (1968). Hemisphere deconnection and unity of consciousness. *American Psychologist*, **23**, 723–33.

Sterkers, G., Pirenne-Ansart, H., Eljaafari-Corbin, A. & Aujard Y. (1993). Towards the identification of the molecular basis responsible for the immaturity of the immune system at birth. Le système immunitaire à la naissance: entre l'apprentissage du soi et du non-soi. *Médecine et Sciences*, **9**, 307–15.

Story, M., Rosenwinkel, K., Himes, J.H., Resnick, M., Harris, L. & Blum, R.W. (1991). Demographic and risk factors associated with chronic dieting in adolescents. *American Journal of Diseases in Childhood*, **145**, 994–8.

Stunkard, A.J., Harris, J.R., Pedersen, N.L., & McClearn, G.E. (1990). The body mass index of twins who have been reared apart. *New England Journal of Medicine*, **322**, 1483–7.

Tanner, J.M. (1986). Use and abuse of growth standards. In *Human growth*, vol 3, ed. F. Falkner & J.M. Tanner, pp. 96–9. New York: Plenum Press.

Tuck, S.J., Monin, P., Duvivier, C., May, T. & Vert, P. (1982). Effect of a rocking bed on apnoea of prematurity. *Archives of Disease in Childhood*, **57**, 475–7.

Tunstall-Pedoe, H. (1988). Autres pays, autres moeurs. Theories on why the French have less heart disease than the British. *British Medical Journal*, **297**, 1559–60.

Underwood, L.E. & D'Ercole, A.J. (1984). Insulin and insulin-like growth factors/somatomedins in fetal and neonatal development. *Clinics in Endocrinology and Metabolism*, **13**, 69–89.
Vague, J. (1956). The degree of masculine differentiation of obesities: a factor determining predisposition to diabetes, arteriosclerosis, gout, and uric calculous diseases. *American Journal of Clinical Nutrition*, **4**, 20–34.
Valoski, A. & Epstein L.H. (1990). Nutrient intake of obese children in a family-based behavioral weight control program. *International Journal of Obesity*, **14**, 667–77.
Van Staveren, W.A., Deurenberg, P., Burema, J., De Groot, L. & Hautvast, J.G. (1986). Seasonal variation in food intake, pattern of physical activity and change in body weight in a group of young adult Dutch women consuming self-selected diets. *International Journal of Obesity*, **10**, 133–45.
Van Vliet, G., Bosson, D., Rummens, E., Robyn, C. & Wolter, R. (1986). Evidence against growth hormone releasing factor deficiency in children with idiopathic obesity. *Acta Endocrinologica (Copenhagen) Supplementum*, **279**, 403–10.
Veldhuis, J.D., Iranmanesh, K.K.Y., Waters, M.J., Johnson, M.L. & Lizarralde, G. (1991). Dual defect in pulsatile growth hormone secretion and clearance subserve the hyposomatotropism of obesity in men. *Journal of Clinical Endocrinology and Metabolism*, **72**, 51–9.
Vignolo, M., Naselli, A., DiBattista, E., Mostert, M. & Aicardi, G. (1988). Growth and development in simple obesity. *European Journal of Pediatrics*, **147**, 242–4.
Wahren, J. (1992). Nutrient induced thermogenesis in health and obesity. *International Journal of Obesity*, **16**(Suppl 1) ix.
Wallace, W.M., Weil, W.B. & Taylor, A. (1958). Effect of variable protein and mineral intake upon the body composition of the growing animal. *CIBA Foundation Colloquia on Ageing*, 4.
Waterlow, J.C. (1974). Some aspects of childhood malnutrition as a public health problem. *British Medical Journal*, **iv**, 88–90.
Waterlow, J.C. & Rutishauser, I.H.E. (1974). Malnutrition in man. In *Early Malnutrition and Mental Development*, ed. J. Cravioto, pp.13–26. Stockholm: Almqvist & Wiksell.
Weiss, G.F., Rogacki, N., Fueg, A., Buhen, D., Suh, J.S., Wong, D.T. & Leibowitz, S.F. (1991). Effect of hypothalamic and peripheral fluoxetine on natural patterns of macronutrient intake in the rat. *Psychopharmacology*, **105**, 467–76.
Widdowson, E.M., Edholm, O.G. & McCance, R.A. (1954). The food intake and energy expenditure of cadets in training. *British Journal of Nutrition*, **8**, 147–55.
Woods, S.C., Decke, E., & Vasselli, J.R. (1974). Metabolic hormones and regulation of body weight. *Psychological Review*, **81**, 26–43.
Wurtman, J. & Baum, M.J. (1980). Estrogen reduces total food and carbohydrate intake but not protein intake in female rats. *Physiology & Behavior*, **24**, 823–7.
Wurtman, R.J., Hefti, F. & Melamed, E. (1981). Precursor control of neurotransmitter synthesis. *Pharmacological Reviews*, **32**, 315–35.
Zapf, J., Walter, H. & Froesch, E.R. (1981). Radioimmunological determination of insulin-like growth factors I and II in normal subjects and in patients with growth disorders and extrapancreatic tumor hypoglycemia. *Journal of Clinical Investigation*, **68**, 1321–30.

Zapf, J., Schoenle, E., Widmer, U. & Froesch, E.R. (1982). Biological effects of the insulin-like growth factors *in vitro*: relevance to their actions *in vivo*. *Bulletin der Schweizerischen Akademie der Medizinischen Wissenschaften*, **98**, 171–82.

Zapf, J., Schmid, C.H. & Froesch, E.R. (1984). Biological and immunological properties of insulin-like growth factors (IGF) I and II. *Clinics in Endocrinology and Metabolism*, **13**, 3–30.

9 *Assessment of body composition in the obese*

E.M.E. POSKITT

Introduction

Definition of obesity

Obesity is an excess of body fat or perhaps more correctly excessive fatness, since it is the proportion of fat to other tissues in the body, rather than the absolute amount of fat, which is indicative of obesity (Bray, 1992). There is no consensus on either the expected amount of fat, or the degree of fatness which can be considered normal. Moreover it is not clear whether it is the total amount of fat in the body, the proportion of fat or the rate at which excess fat is deposited which is most significant for health and fitness. Definition of obesity is complicated by variations in the normal amount and distribution of fat with age, sex, genetic inheritance and physical fitness, irrespective of nutritional state.

Obesity is considered an abnormal situation and one which in current Western culture carries a social disadvantage although fatness is actually considered desirable in some cultures. Severe obesity carries risks for long term health (Hubert *et al.*, 1983; van Itallie & Abraham, 1985; Garrow, 1988) but correlations between culturally undesirable fatness and medically undesirable fatness have not been made. Obesity is difficult to treat and it seems a waste of health resources to struggle with obesity which is of no significance for health. Thus it would be helpful, although currently not possible, to distinguish fatness which is a cosmetic problem only, from fatness which is likely to endanger life.

Fat is the most variable component of the body. It forms less than 10% of body weight in some individuals and more than 50% of body weight in others. Fatness can change noticeably over a few days, often with associated loss of body water and lean body mass. It is not practical to measure the total fat in live individuals directly and there is no 'gold standard' (although estimation of body density by underwater weighing is sometimes considered this), against which to gauge the varied methods of estimating fatness. Thus the value of a method for estimating body composition in obesity is largely determined by factors other than its accuracy. For example, the ability to distinguish rapid changes in lean and fat components may be all-important in slimming and malnourishing states. Rapid loss of fat due to strict dieting is initially accompanied by loss of body water as glycogen stores are utilised (Forbes & Drenick, 1979). There is also loss of lean tissue with loss of fat. Changes in body density in association with slimming will be in part due to loss of body fat and in part due to changes in the composition of lean tissue. If recorded changes in weight or density are interpreted as loss of fat tissue only, the effectiveness of a slimming process will be greatly overestimated.

The distribution of body fat

Fat in the body has two major components: essential and storage fat. Essential fat is part of body structure and is the fat in nervous tissue, bone marrow, cell membranes and that associated with the reproductive organs in females (Gibson, 1990). The remaining fat is usually considered storage fat which waxes and wanes with changes in nutritional status and with other physiological variables. This is fat stored around abdominal and thoracic organs; fat stored retroperitoneally, intra- and intermuscularly; and fat deposited subcutaneously. In normal adults essential fat and storage fat amount to approximately 3% and 12% (males) and 9% and 18% (females) of body weight respectively (Gibson, 1990). About one-third of storage fat is thought to be deposited subcutaneously in normal, non-obese adults (Allen *et al.*, 1956). Shepherd (1991) stated that for every extra 1 kg external fat acquired, an average of 200 g adipose tissue is deposited internally. The relationship between internal and external fat is non-linear and is probably subject to individual variation as well as variation due to age, sex and race (Robson *et al.*, 1971). Very lean individuals are said to have a greater proportion of internal to external fat (Edwards *et al.*, 1955; Allen *et al.*, 1956), but there is some disagreement over this (Skerlj, 1958; Shepherd, 1991). There is also no agreement over whether women have a greater proportion of their fat deposited subcutaneously than men (Garn, 1957; Durnin & Womersley, 1974). Internal fat in children is proportionately less than in adults but shows wide individual variation (Gibson, 1990; Fox *et al.*, 1993).

Storage fat is the fat most susceptible to change, but storage fat in different sites responds differently to metabolic and nutritional changes (Ramirez, 1993). Even storage fat in one part of the body does not behave as one uniform tissue. Fascial planes divide superficial and deeper layers of storage fat. The more superficial layers of fat seem less responsive to nutritional changes than the deeper layers and may have a semistructural rather than storage role (Alexander & Dugdale, 1992).

Variations in body fat with age

Body fatness varies with age. Infants are born with 11–16% body weight (approximately 560 g) as fat (Fomon, 1974; Widdowson, 1974). Most of this fat is deposited in the last trimester of pregnancy. Premature infants are thus less fat than term infants. The infant at 26 weeks of gestation, for example, has about 1% of body weight as fat. This is almost all essential fat (Widdowson, 1974).

In the first months post-term, fat deposition is very rapid. In the time it takes an infant to double birth weight (four to five months) the weight of fat in the body trebles and about 26% of body weight is fat (Fomon, 1974) – predominantly subcutaneous storage fat although deposition of essential fat is also active as myelinisation of the central nervous system takes place. From 6 months until around 5 years old there is a gradual decrease in fatness as body tissue is deposited in greater proportion to fat storage. At 5 years fat is about 12–16% body weight again (Fomon, 1974; Poskitt & Cole, 1977). In the following prepubertal years there is increasing fatness once more (Tanner & Whitehouse, 1975). Boys quite commonly become obese in prepubertal years, particularly when their adolescent growth spurt is delayed. Girls also fatten prepubertally. This 'adiposity rebound' is more likely to develop into persistent obesity if it develops early, i.e. before the age of 6 years (Rolland-Cachera *et al.*, 1984, 1989; Siervogel *et al.*, 1991).

As puberty advances girls have a short period of slimming during their brief growth spurt and then vigorous fat deposition around hips, thighs and breasts leading to the typical adult female fat distribution (Gasser *et al.*, 1993). The growth spurt of late puberty in boys is associated with prolonged deposition of lean tissue. This makes such demands on nutrition that there is not only relatively greater deposition of lean than fat tissue but, often, actual loss of fat as energy to fuel the lean growth spurt (Forbes, 1978). The obesity of male prepubertal years may resolve during the pubertal growth spurt (Tanner & Whitehouse, 1975).

Middle age is commonly associated with increasing fatness. In men this may be a feature of the fourth decade (Gregory *et al.*, 1990). In women fatness may increase following pregnancies or with the onset of the menopause. Although in Western societies such fattening seems almost physiological, 'middle age spread' carries risks for increased morbidity and mortality. This is particularly so when the fat is deposited intra-abdominally. Clinical emphasis on fatness has shifted from concerns about total body fat to concerns about the distribution of body fat. The ability of methods of estimating body fat in the obese to assess or reflect intra-abdominal fat is of considerable clinical importance (van der Kooy & Seidell, 1993).

Estimation of fatness

Estimations of percentage body fat or of fatness tend to refer to the amount or proportion of stored fat as though all methods recorded the same body component. Body fat, fatness, adipose cell number and adipose tissue are not the same. Adipose tissue consists largely of adipose cells which, in obese individuals particularly, are formed predominantly of stored fat. About 80% of adipose tissue is lipid, 15% water and 5% protein (Heymsfield *et al.*, 1992). Adipose cells vary tremendously in size depending on the site from which they are derived and the age and state of nutrition of the individual (Bonnet & Rocour-Brumioul, 1981). Adipose tissue contains supporting stroma of blood vessels and non-adipose connective tissue and, as well as mature adipose cells, variable numbers of adipocyte precursors with, as yet, no stored fat (Widdowson & Shaw, 1973). Adipocytes are largely present in adipose tissue as definite layers or tissue masses internally and externally but they are more diffusely distributed amongst muscle fibres. Adipose tissue is thus more than the total complement of fat-storing adipocytes, and adipocytes are more widely distributed than adipose 'tissue'. Estimates of adipose cell number recognise only those adipose cells containing stored triglyceride. No estimate of body fat, meaning adipose tissue, adipose cells, or stored triglyceride and essential fats, measures any one of these components completely and independently of the other components. Thus it is not surprising that there is little consistency in results of body fat estimates by different methods and that there is no widely accepted measure of obesity (Roche & Chumlea, 1992).

Special considerations relating to measurement of body composition in obesity

The suitability of a method for evaluating body fat in the obese depends on the reasons for the fat estimate. Obesity is a very common problem and *clinical* assessments need to be quick, cheap, repeatable, safe, non invasive and able to show fairly small changes in fatness taking place over short periods of time. The ability of methods precisely to reflect changes in fatness over the course of a few days may be clinically more important than the absolute accuracy of methods. Because of their convenience, low cost, repeatability, safety and subject acceptability, anthropometric methods are the basis of most clinical assessments of obesity. Anthropometric measures can be related to other estimates of fatness derived from smaller experimental groups in order to derive regression equations which are then applied in the clinical situation to give estimates of the fat component of the body from anthropometric data (Durnin & Rahaman, 1967; Brook, 1971; Jackson &

Table 9.1. *Assumptions commonly made when assessing body composition in obesity*

Constancy of composition of adipose tissue
Constancy of composition of lean tissue
Even distribution of excess fat internally and externally
Even distribution of excess fat around the body
No interference of excessive fat with precision of methods

Pollock, 1978, 1980). As with so many results, estimates derived from regression equations are fairly accurate for population estimates but relatively inaccurate for individuals.

Experimental methods of assessing body composition in obesity are discussed elsewhere in this volume, but special considerations are required when dealing with the obese. Body compartment analyses traditionally divide the obese body into fat and lean components. It should not be assumed that with varying fatness, lean body mass remains constant in either quantity or composition. Methods of estimating fatness which involve more than two-compartment modelling are likely to be better at recording the changes in body water and lean body mass accompanying changes in fatness. Concerns about obesity centre on the associated health risks. Clinically useful methods for assessing body composition in obesity should show correlations between estimated obesity and evidence of increased morbidity or mortality. Most methods have not been evaluated in this way.

Table 9.1 lists assumptions often made when estimating total-body fat or percentage fatness. Several are less valid in the obese than in those with a normal range of fatness. Densities of lean and fat tissues, for example, are usually considered constant yet vary throughout the body, and the density of the whole adipose mass may vary according to the degree of fatness and the distribution of adiposity. Water retention is common in obesity and there is often increased extracellular water even in the absence of clinical oedema. This will alter the density of lean tissue and the inferences drawn from estimates of total-body water. Lean body mass is also increased in obesity although we know little about the extent to which this occurs. For every 1 kg increase in weight in obesity, Garrow (1988) estimates 25% of the weight increase is lean tissue. Is the increase in lean tissue only an increase in supporting stroma and blood volume? This depends on the methodology of the fat measurement. Forbes & Welle (1983) estimated that the increase in lean body mass in obesity was 40% of the excess weight. Their measurements estimated lean body mass as body mass minus ether-extractable fat, so their method included the stroma of adipose tissue. Long-term obesity is likely to affect bone density, muscle mass and cardiac size due to the increased weight and blood volume the body has to transport, and these will contribute to the increased lean body mass.

Several methods of assessing obesity, particularly anthropometry and imaging techniques, measure fat at only a few body sites. Yet fat distribution both internally and externally varies widely between individuals and possibly in the same individuals at different ages. Fat is not lost or gained equally in all these areas. Even within the same area there may be different response to fattening and slimming (Alexander & Dugdale, 1992). Changes in one area of adipose tissue should not be assumed to be equivalent to changes elsewhere.

Methods of measuring body composition in obesity

Visual appearance

The eyes are not useful for estimating absolute amounts of body fat but visual assessment is always advisable as part of any nutritional assessment. Looking at individuals can identify and eliminate those with physical abnormalities, such as bony deformities or oedema, which alter some measurements and which can lead to incorrect interpretation of body composition data. For example, indices of weight and height might record an individual who had scoliosis as overweight or obese for height because he or she was recorded as of short stature. Observation should immediately recognise the inappropriateness of such categorisation.

Anthropometry

Weight; weight-for-height; body mass index

To most people, being too fat means being overweight. Weight has many constituents, only one of which is fat. Nevertheless, in the grossly obese excess adipose tissue forms the main contribution to the excess weight.

Obesity has long been judged on the interrelation of weight and height. Independent of the weight of fat, body weight is strongly influenced by height. Many countries (Tanner *et al.*, 1966; van Wieringen, 1972), organisations (WHO, 1983) and some interested groups such as insurance companies (Metropolitan Life Insurance Company, 1959) as well as clothing manufacturers, have developed tables and charts of expected weight-for-height and age. These are useful to the general population as well as to those clinically involved in assessing nutrition. For adults these tables not only relate weight and height but may include 'frame size' which is an attempt to recognise different body builds. In childhood, frame size is often disregarded. One anthropometric assessment of children of the same

Table 9.2. *Weight expected from equivalent centiles as height for boys of the same height but different ages*

	Age (years)	Weight on same centile at same age
Height 109.9 *cm*		
90th centile	4.1	19.7
50th centile	5.0	18.7
10th centile	6.0	17.7
Height 137.5 *cm*		
90th centile	8.7	34.1
50th centile	10.0	31.4
10th centile	11.5	29.7

height but different fatness suggested there was no difference in frame size between fat and lean children (Katch *et al.*, 1991). Those involved in clinical assessment of obese children might, as does this author, question whether this interpretation is valid, since many obese children are not only tall for age but also broad for age. Table 9.2 illustrates the degree to which children who are above average height for age are, on average, heavier at the same height than children who are below average height for age (Forbes, 1972; Amador *et al.*, 1992). Prepubertal children who are obese are usually above average stature (Wolff, 1955; Poskitt, 1980). Thus the interrelation of weight, height and obesity can lead to confusion between overweight and simple tall stature (Poskitt, 1975).

The body mass index (BMI), or (weight, in kg)/(height, in m)^2, developed first by Quetelet (1869), is the most widely used indicator of obesity today. Its popularity is probably helped by the fact that it is one of the few indicators of 'fatness' which has been shown to correlate with mortality risk. The association of mortality risk with BMI follows a U-shaped curve, with increased mortality for those who have very low BMI; average mortality rates for those adults with BMI between 20 and 25 kg/m^2; and a steady increase in mortality risk with those of BMI 28 and above (Garrow, 1988). Mortality rates increase dramatically with BMI over 30 kg/m^2, suggesting this level of BMI is an indication of clinically significant obesity. Whether levels between 25 and 30 kg/m^2, especially those more than 28 kg/m^2, are significant is uncertain.

Garrison & Kannel (1993) looked at cardiovascular risk factors in 2447 non-smoking adults aged 20–59 years. Optimal subscapular skinfolds based on relationship of measurements to these risk factors were <12 mm for men and <15 mm for women. Men with BMI of 22.6 kg/m^2 and women with BMI 21.1 kg/m^2 had a 50% chance of having greater than optimum fatness. The chances of having above-optimum fatness rose rapidly to 90%

of those with BMI >24.5 kg/m^2. The authors felt it was only worth screening for unhealthy adiposity in those with BMI between 22 and 24.5 kg/m^2 (24% of men and 34% women in the reported study). Levels of BMI above 24.5 kg/m^2 could be considered undesirable.

Weight is a reflection of volume and height is a linear measurement. As height increases, weight increases disproportionately. Increased weight requires a body of greater bearing strength. It is likely that those who are overweight because of fat – or for other reasons – must build up their musculoskeletal systems to accommodate the extra weight. BMI is a useful tool for estimating relative fatness in healthy young adults of normal physical fitness. However, where lean body tissue is abnormal in amount or composition, such as with the physically very fit and muscular, or the inactive with very little muscle, BMI loses accuracy. To be exact, it is an indicator of overweight and not of excess fat.

In childhood BMI should not be used by itself as an index of obesity since the relationship between weight and height varies with age and even with height-for-age (Cole, 1991). BMI should be related to standards for age and expressed as relative BMI in centile values (Must *et al.*, 1991; Rolland Cachera *et al.*, 1991), percentage of average (Cole, 1979) or *Z* scores for age (Frisancho, 1990).

The use of relative BMI as a childhood index of obesity has many advantages. It is easy to measure, safe, readily linked into the commonest method of assessing adult obesity (BMI) and has prognostic value in that a tendency to early rebound to increasing BMI indicates a propensity to remain obese into adult life (Rolland Cachera *et al.*, 1984). Many weight-for-height indices have been used to assess obesity in childhood and, as explained earlier, they often have a height bias (Cole, 1979; Rolland-Cachera *et al.*, 1982). There is a great deal of sense in trying to unify methodology thus making studies comparable. Use of a relative BMI would seem a practical way of doing this without the need for expensive equipment. Whatever assessment is used, the precise way 'relative' weight is derived and details of the reference data against which results are compared, should always be stated.

Skinfolds

Skinfold thickness calipers measure folds of subcutaneous tissue which include skin and adipose tissue (Edwards *et al.*, 1955; Durnin & Rahaman, 1967). They only record the thickness of skin and fat at the anatomical sites studied. Skinfolds can be recorded for a great variety of sites on limbs and trunk but practical problems such as subject tolerance usually confine measurements to no more than four or five sites. The popularity of skinfolds for the assessment of obesity is probably confined to those who have *not* tried to use them with the obese. Individuals who are significantly obese frequently have skinfolds fatter than the 40 mm gape of standard calipers. The subcutaneous tissue layer may be so tense in the obese that a fold of skin and fat cannot be raised, or the fold forms a peak which gradually slips through the calipers giving no consistent reading.

The triceps skinfold is probably the commonest skinfold to be used alone. The subscapular skinfold, which is also frequently measured, is more difficult to site consistently. In the obese there are often difficulties raising and measuring a fold on the trunk. Formulae for estimating total body fat have been developed from the relation of combinations of various skinfolds to other estimates of body fat (Brook, 1971; Durnin & Womersley, 1974; Jackson & Pollock, 1978, 1980). These are limited by the problems of measuring skinfolds in the obese and the uncertainties about the relationship of fatfolds to estimates of fatness at the extremes of fatness. Almost certainly the relationship between subcutaneous fat at any site and total body fat does not follow a straight line (Garn, 1957).

Skinfolds only measure external fatness. Their usefulness in defining the degree of adiposity depends on the relationship between fat deposition in subcutaneous and central areas and the relationship between fat in different subcutaneous regions. In children most fat is deposited subcutaneously and skinfolds may provide a fairly good impression of total fatness. In adults the relation of superficial to deep fat varies and correlations between total fat and skinfold thicknesses have been modest. Recent developments have shown the importance of measures of central to peripheral obesity in assessing the health risks of obesity (Vague, 1991; Terry *et al.*, 1992; Croft *et al.*, 1993).

Waist/hip circumference ratio (WHR)

Vague (1956) characterised obesity according to fat distribution as android and gynoid types. The ratio of waist circumference divided by hip circumference (Ashwell *et al.*, 1982) has achieved a great deal of interest in recent years. High WHR in the obese is associated with increased risk of hypertension, diabetes mellitus and coronary heart disease. There is some evidence that the risks of high WHR exist at all levels of fatness (Despres *et al.*, 1988; Raison *et al.*, 1989; Croft *et al.*, 1993).

WHR is a crude measure of fat distribution. Waist measurements will vary with abdominal contents, musculature and both subcutaneous and intra-abdominal fat. More sophisticated techniques of studying fat distribution suggest that it is the intra-abdominal fat which presents the main risk to health (Despres *et al.*, 1988). Waist circumference is now defined as the horizontal circumference in a standing subject, after overnight fast, equidistant between the lower costal margin in the mid-axillary line and the suprailiac crest in the mid-axillary line (Jones *et al.*, 1986), but it has not always been so precisely defined. Thus different studies of WHR may not be totally comparable. Similarly hip circumference measurements have not used a consistent definition of hips until recently. Hip circumference is now defined as the greatest measurement around the buttocks and below the iliac crest (Jones *et al.*, 1986). Other definitions have included a horizontal measure at the level of the greater trochanters. Hip circumference measurements are even more all-embracing than waist measurements since they include not only buttock fatness, but musculature and bony girth. Jakicic *et al.* (1993) found correlations between WHR and lipid profile, controlling for age and BMI, greatest for WHR when 'waist' was halfway between ribs and iliac crest and 'hips' were the widest gluteal circumference. When umbilical circumference was used as 'waist' and 'hip' measurements were as before, there was again a strong association with various blood lipid factors. These associations were found only for those who were obese.

Children have a high WHR compared with mean adult values until puberty, when girls' hips broaden and there may be a lowering of WHR in girls during the later stages of puberty. Boys may or may not retain a high WHR through puberty.

The WHR provides no information on the relative fatness of individuals. It simply provides a non-specific measure of fat distribution. It has been suggested that the ratio provides no more information than is obtainable from waist circumference alone.

Conicity index

There have been other attempts to derive meaningful descriptions of body fat distribution besides waist circumference and WHR. The conicity index (Valdez *et al.*, 1993) uses waist circumference to describe how much greater the abdomen is than might be expected if the body were a perfect cylinder, the assumption being that diversion from a cylindrical shape is due to central adiposity. Thus:

$$\text{Conicity index } (c) = \text{abdominal girth}/(0.109 \text{ weight}/\text{height})$$

Those supporting this index claim its advantages over WHR are a theoretical (expected) range; a built-in waist circumference adjustment for height and weight allowing direct comparison between individuals and populations; and no need for a hip circumference to assess fat distribution. Whilst this index is interesting and does seem to bear some relationship to morbidity and mortality, it does not give any indication of fatness nor of total body fat. As with WHR, it is only an indicator of body fat distribution, not a quantitative assessment of fat.

Body component studies of obesity

Body density

The densities of lean and fat tissue differ. The density of lean tissue is usually given as 1.100 g/ml and that of fat as 0.900 g/ml (Gibson, 1990). Thus, if total body density is known, the relative proportions of fat and lean can be estimated. As with other measures of obesity, formulas relating body density to percentage body fat assume constancy of composition of lean tissue and constancy in the proportions of bone and muscle, lean tissue and water. The standard way of determining body density is by measuring mass in air and then mass during underwater weighing taking account of residual air in the lungs. Currently this method of determining body composition in the obese is taken as the most accurate available for assessing percentage body fat in living subjects. Nevertheless it presents many problems and uses generalisations which are not totally justified in obese individuals (Table 9.1).

Underwater weighing does not appeal to all subjects even though painless. Some subjects are unable to cooperate. The method is not suitable for young children. With the obese there may be problems if subjects are too large to fit into underwater weighing chambers.

Siri's (1961) equation:

$$\text{Percentage body weight as fat} = (495/\text{ density in kg/m}^3) - 450$$

is widely used to determine a figure for percentage body weight as fat from estimations of density. The composition, and thus the density, of lean tissue is assumed to be constant after the age of about 3 years. Young children and, particularly, premature infants have much higher total body water, especially extracellular water, and this lowers the density of lean tissue (Widdowson, 1974). Lean tissue from individuals of different races may not have the same density, presumably because of different proportions of muscle and bone to other tissues. Water retention is common in the severely obese. Total-body water is higher than in normal, not overfat individuals. Deurenberg *et al.* (1989*a*) argued that density of fat-free mass decreases with increasing fatness and can lead to 2–4% overestimation of the percentage of body fat, depending on the severity of the obesity. Adipose tissue also varies in density. This may be due to variations in the size of fat cells and in the amounts of supporting stroma. Gibson (1990) has summarised some of the problems in estimating body fat from measurements of density. Errors in normally fat individuals may arise from variations in bone density leading to approximately 2% variation in body fat. In the obese it is likely that variations of this kind will lead to even greater inaccuracies.

Total-body water

Body water is contained in the lean tissue compartment of the body. Lean tissue is assumed to contain 73.2% water. Estimation of body composition from a two-compartment model uses total-body water measurements determined by the hydrogen isotopes deuterium or tritium, or the stable oxygen isotope ^{18}O, to estimate lean mass. Fat mass, or percentage body weight as fat, can then be determined by subtraction from body weight. Body water estimations will again be influenced by the increase in extracellular water common in obesity. Body fat will be underestimated. Methods and interpretation in young children need to take account of the increased total body water of the young and the different distribution between extra- and intracellular spaces in the young.

Total-body potassium

Natural gamma rays from ^{40}K are emitted by the body. Measurement of gamma ray emission is indicative of total-body potassium, which is itself indicative of the body cell mass, that is the metabolically active non-fat tissue or lean body mass excluding extracellular water, cartilage and bone and ligaments. Again interpretation of lean mass and thus fat mass assumes constancy in the composition of lean body mass.

Measurement of total-body potassium in the gamma counter is safe and painless but it is a slow, expensive process. Total-body potassium is estimated by comparing emission rates with those from phantoms formed of cylinders filled with a potassium-containing solution of known composition and volume, placed so as to mimic the subject's shape and size. For obese individuals, the chamber for measuring emission may seem claustrophobic and too small. For children phantoms have to be designed that are appropriate to age and height. ^{40}K emission is often underestimated in the obese because fat tissue tends to absorb gamma rays to a small extent. Lean body mass is thus underestimated and obesity overestimated in those who are already significantly obese.

Bioelectrical impedance (BEI)

Fat and fat-free mass have different electric conductivities. The impedance of an electric current is proportional to the length of the conductor and inversely proportional to the sectional area. Formulas which use these principles can be derived to determine body density from conductance or impedance. BEI has become a popular method for determining the proportions of lean and fat in the body. It is simple, painless, and produces results which look impressive as a print-out. Using a single frequency (usually 50 kHz) impedance is not very satisfactory for demonstrating individual changes during weight loss (Tagliabue *et al.*, 1992). Changes in BEI are largely due to changes in body water distribution (Deurenberg *et al.*, 1989*c*). Problems therefore arise when BEI is used to attempt to follow changes in lean and fat mass with slimming. Tagliabue *et al.* (1992) found very variable results in estimated fat-free mass in 27 subjects slimming successfully over 10 weeks. The estimate of the change in fat-free mass in one individual was even greater than the actual weight loss. The appeal of the method and the ease with which results can be obtained may obscure an accuracy which is no greater than that of the best anthropometry (Deurenberg *et al.*, 1989*b*) and which may totally fail to document changes in body fat over time. Using different frequencies may be more complicated but probably distinguishes differences between extra- and intracellular water.

Dual-energy X-ray absorptiometry (DXA)

X-rays of different energies are attenuated to different extents by fat and lean tissues. If no bone is present, measures of attenuation relate to lean and fat content. Where bone is present, results can be extrapolated. The dose of X-rays is very small, but even so this should discourage frequent use in the same individuals. The technique is not useful in measuring intra-abdominal fat directly and the advantages of a fairly specialised technology which cannot distinguish subcutaneous and intra-abdominal fat seem few.

Near-infrared spectrophotometry (NIR)

The methodology of NIR has been widely publicised as part of a commercial venture. Equipment which measures re-emission light from a ‘lightwand’ emitting low-level electromagnetic resonance at 940 nm and 950 nm over the biceps skinfold produces an estimate of body composition using programs which incorporate thickness of the skinfold, weight, height and age. Hortobagyi *et al.* (1992) found NIR significantly underestimated the percentage of body fat. Optical density readings accounted for less than 46% of the variation in skinfold thickness. Regression models showed that optical density readings did not predict any additional significant variance in percentage body fat compared with other social and anthropometric data fed into the regression equation. There seems little evidence that the method is producing results which could not be achieved by recording weight, height and age alone. For individuals, results are very inaccurate.

Body imaging techniques

Ultrasound

Ultrasound measurement of the thickness of the subcutaneous fat layer is painless, non-invasive and relatively cheap. The technique is simple and requires little training to perform. It provides more accurate measurement of skinfold thickness than calipers and is useful where skinfolds are too thick to be measured by calipers. It is particularly useful for detecting changes in fat and muscle layers in the limbs (Weits *et al.*, 1986).

Ultrasonography can be used to obtain an estimate of intra-abdominal fat by measuring intra-abdominal thickness from rectal muscle to aorta. This measure seems to correlate quite well with computed tomography estimations of internal fat (Armellini *et al.*, 1993).

Computed tomography (CT)

CT can be used to provide cross-sectional views of any part of the body. It involves radiation and thus is not appropriate when repeated views are needed. It is very expensive, quite slow to perform if more than one scanning slice is used and thus of limited use. If total body fat is being measured multiple slices are needed, involving significant radiation. The technique is one of the few which can demonstrate internal fat and its main use in obesity is to demonstrate the proportion of intra-abdominal fat to fat in the abdominal wall. Single scans taken at the level of L4–5 (approximately the level of the umbilicus) are highly correlated with total visceral fat (Rossner *et al.*, 1990).

The accuracy of CT scanning for defining intra-abdominal fat is affected by a 'partial volume' effect causing misclassification of some pixels. This is particularly likely for intra-abdominal fat because of the alignment of fat with the irregular border of the intestines. Although intestinal fat can be distinguished, differentiation of intraperitoneal fat from peritoneal fat is not possible. Ashwell *et al.* (1987) defined fat behind the mid-point of the abdominal aorta and inferior vena cava as retroperitoneal fat.

Kvist *et al.* (1988*a*) have developed equations for estimation of total-body fat from 22 scans. The value of these and other equations needs further evaluation. Comparison of total-body fat estimated in this way and weight/height indices suggested weight/height$^{0.9}$ was the best anthropometric index for the estimation of total-body fat (Kvist *et al.*, 1988*b*).

Table 9.3. *Advantages and disadvantages of some methods of assessing obesity*

Method	Cost	Measures body fat	Measures IA fat	General comments
BMI	Low	No	No	Simple; repeatable; safe; some predictive value
Skinfolds	Low	Limited sites only	No	Cannot be measured in the very obese
WHR	Low	No	Partly	Has been poorly defined; some predictive value
Underwater weighing	High	No	No	Widely regarded as most accurate method for total body fat
Ultrasound	Moderate	No	Partly	Most practical method for IA fat
CT	Very high	Possible	Yes	Most available method for good IA fat estimation; involves radiation
MRI	Very high	Possible	Yes	Not widely available; prohibitive cost

IA fat, intra-abdominal fat; BMI, body mass index; WHR, waist/hip circumference ratio; CT, computed tomography; MRI, magnetic resonance imaging.

Magnetic resonance imaging (MRI)

MRI is even more expensive than CT and is a slower procedure. It does not involve radiation and could thus be used for serial measurements were it not so expensive and if it were more readily available. It produces excellent images of body fat distribution. Estimates of total-body fat can be made by summing the areas of fat on numerous cross-sectional slices. MRI is also useful for demonstrating the distribution of internal to external fat. It is likely to remain an experimental rather than a clinical procedure for demonstrating body composition in obesity.

Conclusions

All methods of assessing body composition in the obese are flawed (Table 9.3). The limitations which apply to body composition estimation in all individuals are no less relevant in the obese. In many cases additional problems are posed by alterations in lean body mass for height, total-body water, and by the sheer size and fatness of some obese individuals.

Obesity remains a common clinical problem. In clinical practice two anthropometric measures only relate to prognosis: BMI and WHR. This is despite widespread use of bioelectrical impedance within some clinical situations. Intra-abdominal fat seems to be the important fat in terms of risk of morbidity and mortality from obesity-related disorders. Although it can be estimated using ultrasound, intra-abdominal fat cannot be measured with any precision except by expensive imaging techniques. Preferences for some of the more sophisticated, invasive, expensive or experimental methodologies will vary with the facilities and/or skills of researchers. Nevertheless, the methods of estimating body composition in the obese are, despite vigorous attempts to apply new methodologies and define acceptable technologies, still largely unsatisfactory.

References

Alexander, H.G. & Dugdale, A.E. (1992). Fascial planes within subcutaneous fat in humans. *European Journal of Clinical Nutrition*, **46**, 903–6.

Allen, T.H., Peng, M.T., Chen, K.P., Huang, T.F., Chang, C. & Fang, H.S. (1956). Prediction of total adiposity from skinfold and the curvilinear relationship between external and internal adiposity. *Metabolism*, **5**, 346–352.

Amador, M., Bacallao, J. & Hermelo, M. (1992). Adiposity and growth: relationship of stature at fourteen years with relative body weight at different ages and several measures of adiposity and body bulk. *European Journal of Clinical Nutrition*, **46**, 213–19.

Armellini, F., Zamboni, M., Robbi, R. *et al.* (1993). Total and intra-abdominal fat measurements by ultrasound and computerised tomography *International*

Journal of Obesity, **17**, 209–14.
Ashwell, M., Chinn, S. & Garrow, J.S. (1982). Female fat distribution – a simple classification based on two circumference measurements. *International Journal of Obesity*, **6**, 143–52.
Ashwell, M., McCall, S.A., Cole, T.J. & Dixon, A.K. (1987). Fat distribution and its metabolic consequences: interpretations. In *Human Body Composition and Fat Distribution*, ed. N.G. Norgan, pp 227–43. The Hague: CIP gegevens Koninkijke Bibliothek.
Bonnet, F.B. & Rocour-Brumioul, D. (1981). Normal growth of human adipose tissue. In *Adipose Tissue in Children*, ed. F.P. Bonnet, pp. 81–107. Boca Raton: CRC Press.
Bray, G.A. (1992). An approach to the classification and evaluation of obesity. In *Obesity*, ed. P. Bjorntorp & B.N. Brodoff, pp. 294–308. Philadelphia: Lippincott.
Brook, C.G.D. (1971). Determination of body composition in children from skinfold measurements. *Archives of Disease in Childhood*, **48**, 725–8.
Cole, T.J. (1979). A method for assessing age standardised weight-for-height in children seen cross-sectionally. *Annals of Human Biology*, **7**, 457–473.
Cole, T.J. (1991). Weight-stature indices to measure underweight, overweight and obesity. In *Anthropometric Assessment of Nutritional Status*, ed. J.H. Himes, pp. 83–111. New York: Wiley-Liss.
Croft, J.B., Strogatz, D.S., Keenan, N.L. James, S.A., Malarcher, A.M. & Garrett, J.M. (1993). The independent effects of obesity and body fat distribution on blood pressure in black adults: the Pitt County Study. *International Journal of Obesity*, **17**, 391–8.
Despres, J-P., Tremblay, A., Perusse, L., Leblanc, C. & Bouchard, C. (1988). Abdominal adipose tissue and serum HDL-cholesterol: association independent from obesity and serum triglyceride concentration. *International Journal of Obesity*, **12**, 1–13.
Deurenberg, P., Leenan, K., van der Kooy, K. & Hautvast, J.G.A.T. (1989*a*). In obese subjects the body fat percentage calculated with Siri's formula is an overestimate. *European Journal of Clinical Nutrition*, **43**, 569–75.
Deurenberg, P., Smit, H.E. & Kusters, C.S.L. (1989*b*). Is the bioelectrical impedance method suitable for epidemiological field studies? *European Journal of Clinical Nutrition*, **43**, 647–54.
Deurenberg, P., van der Kooy, K., Leenan, R. & Schouten, F.J.M. (1989*c*). Body impedance is largely dependent on the intra- and extra-cellular water distribution. *European Journal of Clinical Nutrition*, **43**, 845–53.
Durnin, J.V.G.A. & Rahaman, M.M. (1967). The assessment of the amount of fat in the human body from measurements of skinfold thickness. *British Journal of Nutrition*, **21**, 681–9.
Durnin, J.V.G.A. & Womersley, J. (1974). Body fat assessed from total body density and its estimation from skinfold thickness: measurements on 481 men and women from 16–72 years. *British Journal of Nutrition*, **32**, 77–97.
Edwards, D.A., Hammond, W.H., Heely, M.J.R., Tanner, J.M. & Whitehouse, R.H. (1955). Design and accuracy of calipers for measuring subcutaneous tissue thickness. *British Journal of Nutrition*, **12**, 133–43.
Fomon, S.J. (1974). *Infant Nutrition* 2nd edn., p. 69. Philadelphia: Saunders.

Forbes, G.B. (1972). Relation of lean body mass to height in children and adolescents. *Pediatric Research*, **6**, 32–7.

Forbes, G.B. (1978). Body composition in adolescence. In *Human Growth*, vol. 2, ed. F. Falkner & J.M. Tanner, pp. 239–72. London: Ballière-Tindall.

Forbes, G.B. & Drenick, E. J. (1979). Loss of body nitrogen on fasting. *American Journal of Clinical Nutrition*, **32**, 1570–4.

Forbes, G.B. & Welle, S.L. (1983). Lean body mass in obesity. *International Journal of Obesity*, **7**, 99–107.

Fox, K., Peters, D., Armstrong, N., Sharpe, P. & Bell, M. (1993). Abdominal fat deposition in 11-year-old children. *International Journal of Obesity*, **17**, 11–16.

Frisancho, A.R. (1990). *Anthropometric Standards for the Assessment of Growth and Nutritional Status*, p. 43. Ann Arbor: University of Michigan Press.

Garn, S.M. (1957). Fat weight and fat placement in the female. *Science*, **12**, 1092–3.

Garrison, R.J. & Kannel, W.B. (1993). A new approach for estimating healthy body weights. *International Journal of Obesity*, **17**, 417–23.

Garrow, J.S. (1988). *Obesity and Related Diseases*, pp. 1–19. Edinburgh: Churchill Livingstone.

Gasser, T., Ziegler, P., Kneip, A., Prader, A., Molinari, L. & Largo, R.H. (1993). The dynamics of growth of waist circumference and skinfolds in distance velocity and acceleration. *Annals of Human Biology*, **20**, 239–59.

Gibson, R. (1990). *Principles of Nutritional Assessment*, pp. 187–208. Oxford: Oxford University Press.

Gregory, J., Foster, K., Tyler, H. & Wiseman, M. (1990). *The Dietary and Nutritional Survey of British Adults*, p. 232. London: HMSO.

Heymsfield, S.B., Lichtman, S., Baumgartner, R.N., Dilmanian, F.A. & Kamen, F. (1992). Assessment of body composition: an overview. In *Obesity*, ed. P. Bjorntorp & B.N. Brodoff, pp. 37–54. Philadelphia: Lippincott.

Hortobagyi, T., Israel, R.G., Houmard, J.A., McCannon, M.R. & O'Brien, K.F. (1992). Comparison of body composition assessment by hydrodensitometry, skinfolds and multiple site near infra-red spectrophotometry. *European Journal of Clinical Nutrition*, **46**, 205–11.

Hubert, H.B., Feinleib, M., McNamara, P.M. *et al.* (1983). Obesity is an independent risk factor for cardiovascular disease. A 26 year follow-up of participants in the Framingham Heart Study. *Circulation*, **67**, 968.

Jackson, A.S. & Pollock, M.L. (1978). Generalised equations for predicting body density of men. *British Journal of Nutrition*, **40**, 497–504.

Jackson, A.S., Pollock, M.L. & Ward, A. (1980). Generalised equations for predicting body density of women. *Medical Science of Sports and Exercise*, **12**, 175–83.

Jakicic, J.M., Donnelly, J.E., Jawad, A.F., Jacobsen, D.J., Gunderson, S.C. & Pascale, R. (1993). Association between blood lipids and different measurements of body fat distribution: effects of body mass index and age. *International Journal of Obesity*, **17**, 131–7.

Jones, P.R.M., Hunt, M.J., Brown, T.P. & Norgan, N.G. (1986). Waist-hip circumferences ratio and its relation to age and overweight in British men. *Human Nutrition: Clinical Nutrition*, **40C**, 239–247.

Katch, V., Becque, M.D., Marks, C., Moorehead, C. & Rocchini, A. (1991). Gender dimorphism in size, shape and body composition of child-onset obese and non-obese adolescents. *International Journal of Obesity*, **15**, 267–82.

Kvist, H., Chowdhury, B., Grangard, U., Tylen, U. & Sjostrom, L. (1988*a*). Total and visceral adipose tissue volumes derived from measurements with computerised tomography in adult men and women: predictive equations. *American Journal of Clinical Nutrition*, **48**, 1351–61.

Kvist, H., Chowdhury, B., Sjostrom, L., Tylen, U. & Cederblad, A. (1988*b*). Adipose tissue volume determination in males by computerised tomography and ^{40}K. *International Journal of Obesity*, **12**, 246–66.

Metropolitan Life Insurance Company. (1959). New weight standards for men and women. *Statistics Bulletin of Metropolitan Life Company*, **40**, 1–10.

Must, A., Dallal, G.E. & Dietz, W.H. (1991). Reference data for obesity: 85th and 95th percentiles of body mass index (W/H^2) – a correction. *American Journal of Clinical Nutrition*, **54**, 773.

Poskitt, E.M.E. (1975). Defining malnutrition in the young child. Lancet, **ii**, 348.

Poskitt, E.M.E. (1980). Obese from infancy: a re-evaluation. In *Nutrition*. Topics in Paediatrics 2, ed. B.A. Wharton, pp. 81–9. Bath: Pitman Medical.

Poskitt, E. M. E. & Cole, T. J. (1977). Do fat babies stay fat? *British Medical Journal*, **i**, 7–9

Quetelet, L.A.J. *Physique Sociale*, vol 2, p. 92. Brussels: Muquardt.

Raison, J., Bonithon-Kopp, C., Guy-Grand, B. & Ducimetriere, P. (1989). Body fat distribution and metabolic parameters in a healthy French female population in comparison with obese women. In *Obesity in Europe*, ed. P. Bjorntorp & S. Rossner, pp. 43–8. London: Libbey.

Ramirez, M.E. (1993). Familial aggregates of subcutaneous fat deposits and the peripheral fat distribution pattern. *International Journal of Obesity*, **17**, 63–8.

Robson, J.R.F., Bazin, M. & Soderstrom, R. (1971). Ethnic differences in skinfold thickness. *American Journal of Clinical Nutrition*, **24**, 864–8.

Roche, A.F. & Chumlea, W.M.C. (1992). New approaches to the clinical assessment of adipose tissue. In *Obesity*, ed. P. Bjorntorp & B.N. Brodoff, pp. 55–66. Philadelphia: Lippincott.

Rolland-Cachera, M-F., Sempé, M., Guilloud-Bataille, M., Patois, E., Pequignot-Guggenbuhl, F. & Fautrad, V. (1982). Adiposity indices in children. *American Journal of Clinical Nutrition*, **36**, 176–84.

Rolland-Cachera, M-F., Deheeger, M., Bellisle, F., Sempé, M., Guilloud-Bataille, M. & Patois, E. (1984). Adiposity rebound in children: a simple indicator for predicting obesity. *American Journal of Clinical Nutrition*, **39**, 129–35.

Rolland-Cachera, M-F., Bellisle, F. & Sempé, M. (1989). The prediction in boys and girls of the weight/height^2 index and various skinfold measurements in adults: a two decade follow-up study. *International Journal of Obesity*, **13**, 305–11.

Rolland-Cachera, M-F., Cole, T.J., Sempé, M., Tichet, J., Rossignol, C. & Charraud, A. (1991). Body mass index variations: centiles from birth to 87 years. *European Journal of Clinical Nutrition*, **45**, 13–21.

Rossner, S., Bo, W.J., Hiltbrandt, E. *et al.* (1990). Adipose tissue determination in cadavers – a comparison between cross-sectional planimetry and computerised tomography. *International Journal of Obesity*, **14**, 893–902.

Shepherd, R. (1991). *Body Composition in Biological Anthropometry*, p. 64.

Cambridge: Cambridge University Press.

Siervogel, R.M., Roche, A.F., Guo, S., Mukherjee, D. & Chumlea, W.C. (1991). Patterns of change in weight/stature^2 from 2 to 18 years: findings from long-term serial data for adults in the Fels longtitudinal growth study. *International Journal of Obesity*, **15**, 479–85.

Skerlj, B. (1958). Age changes in fat distribution in the female body. *Acta Anatomica*, **38**, 56–63.

Tagliabue, A., Cena, H., Trentani, C., Lanzola, E. & Silva, S. (1992). How reliable is bioelectrical impedance analysis for individual patients? *International Journal of Obesity*, **16**, 649–52.

Tanner, J.M. & Whitehouse, R.H. (1975). Revised standards for triceps and subscapular skinfolds in British children. *Archives of Disease in Childhood*, **50**, 142–5.

Tanner, J.M., Whitehouse, R.H. & Takaishi, M. (1966). Standards from birth to maturity for height, weight, height velocity, and weight velocity: British children 1965. *Archives of Disease in Childhood*, **41**, 454–71, 613–35.

Terry, R.B., Page, W.F. & Haskell, W.L. (1992). Waist-hip ratio, body mass index and premature cardiovascular disease mortality in US Army veterans during a twenty three year follow-up study. *International Journal of Obesity*, **16**, 417–24.

Vague, J. (1956). The degree of masculine differentiation of obesity: a factor determining predisposition to diabetes, gout and uric calculous disease. *American Journal of Clinical Nutrition*, **4**, 20–34.

Vague, J. (1991). *Obesities*, pp. 41–58. London: Libbey.

Valdez, R., Seidell, J.C., Young, I.A. & Weiss, K.M. (1993). A new index of abdominal adiposity as an indicator of risk for cardiovascular disease. A cross population study. *International Journal of Obesity*, **17**, 77–82.

van der Kooy, K. & Seidell, J.C. (1993). Techniques for the measurement of visceral fat: a practical guide. *International Journal of Obesity*, **17**, 187–96.

van Itallie, T. B. & Abraham, S. (1985). Some hazards of obesity and its treatment. In *Recent Advances in Obesity Research* IV, ed. J. Hirsch & T.B. van Itallie, pp. 1–19. London: John Libbey.

van Wieringen, J. C. (1972). *Seculaire Groeiverschuiving*. Leiden: Nederlands Institut voor Praeventieve Geneeskunde.

Weits, T., vander Beek, E. J. & Wedel, M. (1986). Comparison of ultrasound and skinfold caliper measurement of subcutaneous fat tissue. *International Journal of Obesity*, **10**, 161–8.

Widdowson, E. M. (1974). Changes in body proportions and body composition during growth. In *Scientific Foundations of Paediatrics*, ed. J.A. Davis & J. Dobbing, pp. 153–63. London: Heinemann.

Widdowson, E. M. & Shaw, W. J. (1973). Full and empty fat cells. *Lancet*, **ii**, 905.

Wolff, O. H. (1955). Obesity in childhood. A study of birthweight, height and onset of puberty. *Quarterly Journal of Medicine*, **24**, 109–23.

World Health Organization. (1983). *Measuring Change in Nutritional Status*. Geneva: WHO.

10 *The role of body physique assessment in sports science*

MARY H. SLAUGHTER AND CONSTANCE B. CHRIST

Introduction

Physique is comprised of three distinct, yet interrelated components, namely, body size, structure and composition (Boileau & Lohman, 1977). The elements of physique thus defined have been adopted for this text. Body size refers to the physical magnitude of the body and its segments (e.g. length, mass, volume, surface area). Body structure alludes to the distribution or body parts expressed as ratios (e.g. inverse ponderal index). Body composition adverts to the amount of the various constituents in the body (e.g. fat, water, mineral, protein).

Interest in the influence of body physique on physical performance dates back to antiquity (Gunther, 1975). Contrast the physique of a sumo wrestler with that of a marathon runner, or the physique of a volleyball player with that of a gymnast. Two striking observations become immediately apparent. Within a particular sport and position, the physiques of elite athletes appear similar. In contrast, body size, structure and composition differ markedly among athletes of different sports.

Why is the assessment of physique important? The assessment of physique can be used to ascertain an individual's readiness for sport participation in terms of injury prevention. A chemically immature skeleton is less able to withstand the trauma and shearing stresses incurred in contact sports (e.g. football) or in sports with repeated internal trauma (e.g. long distance running, pitching) than a mature skeleton, thus rendering the athlete more prone to fractures and avulsions. The assessment of physique can also be utilised to characterise the profiles of athletes in different sports. An individual with a particular profile can be encouraged to participate in the sport for which he or she is physically best suited. This, in turn, may make sport participation a more rewarding experience by increasing the likelihood of success and decreasing the likelihood of injury. In addition, the assessment of physique can be used to determine optimal weight for optimal health and performance. Theoretically, a mechanical advantage can be gained in weight class sports (e.g. wrestling, judo) by qualifying for the next-lowest weight class. Athletes are often pressured to lose weight in order to compete in a lower weight class. If optimal weight (including optimal fatness) is known, the deleterious practice of acute weight reduction to meet lower weight class requirements can be ameliorated. Similarly, prevalent eating disorders in sports in which body image is important (e.g. dance, figure skating, gymnastics, diving) can be curtailed. Further, the potential health risks associated with low body fat (e.g. amenorrhoea and compromised skeletal health) in many female athletes can be diminished.

How does physique influence performance? To answer this question one must consider dimensional analysis and the theory of biological similarity which purports that changes in body dimensions follow geometric relationships (Thompson, 1917; Asmussen & Heeboll-Nielsen, 1955; Asmussen, 1974; Gunther, 1975). If a linear dimension (e.g. stature) is altered by a certain percentage, all other linear dimensions (e.g. trunk length, arm length) will be altered by the same percentage. An area is proportional to the square of a linear dimension (L^2), whereas volume or mass is proportional to a linear dimension cubed (L^3). The impact of physique on physical performance becomes apparent when these principles are applied to two individuals who differ in stature. Suppose subject B is 50% taller than subject A (i.e. A = 1, B = 1.5). Theoretically, the length (L), area (L^2) and volume (L^3) constants in subject A would equal 1, 1^2, and 1^3, respectively, whereas the length, area and volume constants in subject B would equal 1, 2.25 and 3.375, respectively.

Dimensional analysis can be applied to determine whether subject A or subject B will be able to perform more work if they differ only in stature. Work = force × distance. Since force is proportional to the cross-sectional area of the muscle exerting the force (L^2) and distance is a linear dimension (L), work is proportional to L^3. Therefore, a taller stature is advantageous for doing work and for exerting force against an external object. In contrast, when the body must be moved against gravity a shorter stature is favoured. For example, the mechanics underlying chin-ups can be reduced to: (force × force arm)/(resistance × resistance arm), or $(L^2 \times L)/(L^3 \times L)$, which is equal to L^{-1}. Similarly, a shorter stature is favoured for acceleration (i.e. sprinting) where acceleration = force/mass, or $L^2/L^3 = L^{-1}$. Therefore, subject A would fare better in accelerating or in lifting his weight against gravity (e.g. gymnastics, track events); whereas subject B would be able to exert more force (American football, baseball, shot put) and, therefore, do more work.

Measurement techniques

The most common measures of body size are stature and weight. Somatotype has been utilised to depict body structure (Sheldon *et al.*, 1954; Parnell,

1958; Heath & Carter, 1967; Carter & Heath, 1990). Numerous ratios have also been employed to characterise the distribution or arrangement of body parts. For example, inverse ponderal index (height/weight$^{\frac{1}{3}}$) has been used as an index of the massiveness of the body, with a lower number reflecting a more ponderous form and a larger number reflecting a more linear shape (Sheldon, 1940). The Heath–Carter method of somatotyping also employs the use of the inverse ponderal index to assess the degree of ectomorphy (Heath & Carter, 1967). The somatotype approach, however, does not distinguish fat from lean mass.

Body mass index (weight/height^2) (Keys *et al.*, 1972; Garrow & Webster, 1985), and waist circumference/hip circumference (Seidell, 1991) have been used as indices of obesity. Body mass index (BMI) is highly correlated with weight and with fat as determined by densitometry (Keys *et al.*, 1972). However, the use of BMI as a marker of obesity has been criticised for three reasons (Garn *et al.*, 1986). First, BMI is not completely independent of stature, particularly in younger individuals. Moreover, BMI may be affected by relative leg length or sitting height, as short-leggedness yields higher BMI values. Further, the numerator (weight) does not distinguish fat from lean mass; hence, the numerator can theoretically reflect both fat and lean tissue comparably. Waist circumference/hip circumference has been used as an indicator of morbidity, metabolic disorders, cardiovascular disease, stroke, diabetes mellitus and total mortality. Unfortunately, the measurement techniques for the two circumferences are not well standardised, and the circumferences do not distinguish intra-abdominal fat from subcutaneous fat.

Fat-free mass index (fat-free mass/height^2) has been used as a marker of nutritional status (Van Itallie et al., 1990). Shoulder width/hip width has been used to indicate the degree of athleticism or mesomorphy, as the shoulder width in athletes typically exceeds their hip width. Sitting height/standing height has been used to assess leg length. Additional information regarding the use of ratios to depict body structure can be found in Fidanza (1991).

A limitation in utilising ratios to assess body structure is that a specific relationship is assumed: a straight line with the slope equal to the mean fat-free body (FFB) over the mean height and an intercept of zero. Both theoretical and empirical evidence has indicated that these assumptions are tenuous. To circumvent this problem, Benn (1971) employed the use of a regression model in which he retrospectively determined the weight-to-height^x relationship and used it as a marker of adiposity. More recently, the regression of FFB on height^x has been applied as an objective method to assess musculoskeletal size in both athletic and non-athletic populations (Slaughter & Lohman, 1980; Slaughter *et al.*, 1987).

Numerous measurement techniques have been used to assess body composition, including anthropometry, densitometry, hydrometry, gamma ray spectrometry, bioelectrical impedance, total-body electrical conductivity, dual-energy X-ray absorptiometry, computed tomography, magnetic resonance imaging and neutron activation. Reviews of these techniques are provided elsewhere (Norgan, 1985, 1991; Lohman, 1986, 1992; Forbes, 1987; Brodie, 1988*a*, *b*; Bosaeus & Isaksson, 1991; Heymsfield, 1991).

Densitometry has been generally accepted as the criterion method used to determine body composition; it is the technique against which all others are validated. Body density is affected by the relative contribution of the individual components which comprise the FFB. Therefore, unaccounted-for variability in the water, mineral or protein content of the FFB will compromise the accuracy of body composition estimates from densitometry. Maturational status (Boileau *et al.*, 1984; Lohman *et al.*, 1984*b*), age (Haschke *et al.*, 1981; Fomon *et al.*, 1982; Boileau *et al.*, 1992), gender (Fomon *et al.*, 1982; Lohman *et al.*, 1984*b*), race (Lohman *et al.*, 1984*b*; Slaughter *et al.*, 1990) and participation in sports (Bunt *et al.*, 1990) alter the relative amounts of the subcomponents of the FFB, and therefore FFB density. Hence, two individuals with identical percentage body fat but different relative distributions of the subcomponents of the FFB will have different body densities, and consequently different estimates of percentage fat. For a numerical example, see Lohman *et al.* (1984*a*). Meaningful interpretation of body composition data is contingent upon an understanding of the assumptions underlying the measurement technique used to acquire the data, and whether any of those assumptions has been violated in the particular sample measured. If, for example, the assumption of a constant FFB density is violated, then a model which accounts for the variability in water, mineral and protein content should be employed in order to obtain accurate estimates of body composition (Boileau *et al.*, 1985; Lohman, 1986).

Body composition of athletes

Estimates of body composition typically divide the body into fat and fat-free (water, mineral, protein) components. Adult athletes in all sports are typically leaner and have more FFB than their non-athletic counterparts, with few exceptions. However, the precise fat and fat-free masses observed in athletes are contingent upon gender, level of competition, and the specific sport considered. Of the male athletes examined, the mean FFB ranged from 65 kg for the cyclists to 82 kg for the judoists (Fig. 10.1). In females, the FFB ranged from 43 kg for the gymnasts to 57 kg for the basketball players (Fig. 10.2). Percentage fat for the male athletes fell within a narrow range: 7% for the distance runners to about 11% for the Olympic lifters (Fig.

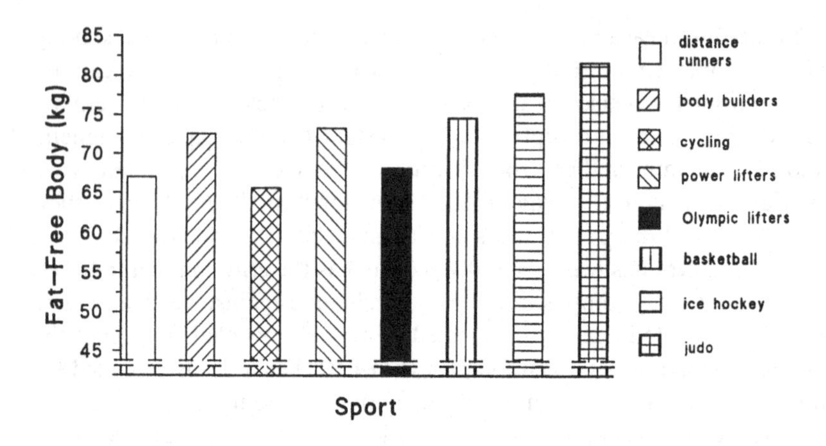

Fig. 10.1. Mean fat-free body in male athletes by sport.

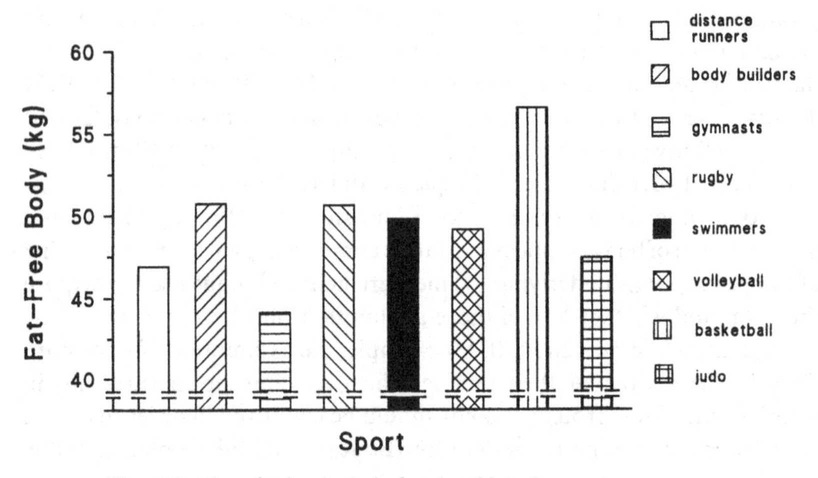

Fig. 10.2. Mean fat-free body in female athletes by sport.

10.3). In contrast, greater variability in percentage fat was observed among female athletes: 16% for the body builders and judo participants to 22% for the swimmers and basketball players (Fig. 10.4).

Caution, however, must be exerted in interpreting these results. First, these data were derived from densitometry, and a multicomponent model to account for variability in water, mineral or protein content was not employed. There is evidence that athletes deviate from reference man with the site and magnitude of the deviation being dependent upon gender, sport and level of competition. Contrast, for example, reference man with an American football lineman and a female long distance runner. It is likely that the FFB density of the football lineman will be greater than reference

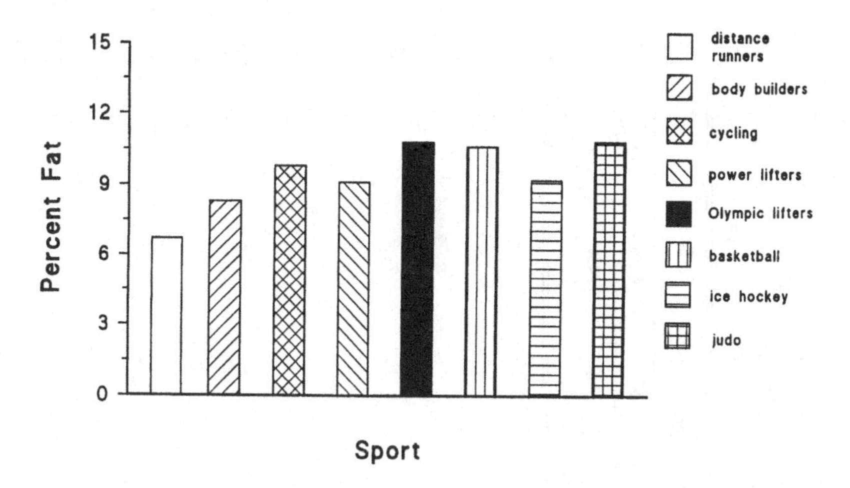

Fig. 10.3. Mean percentage fat in male athletes by sport.

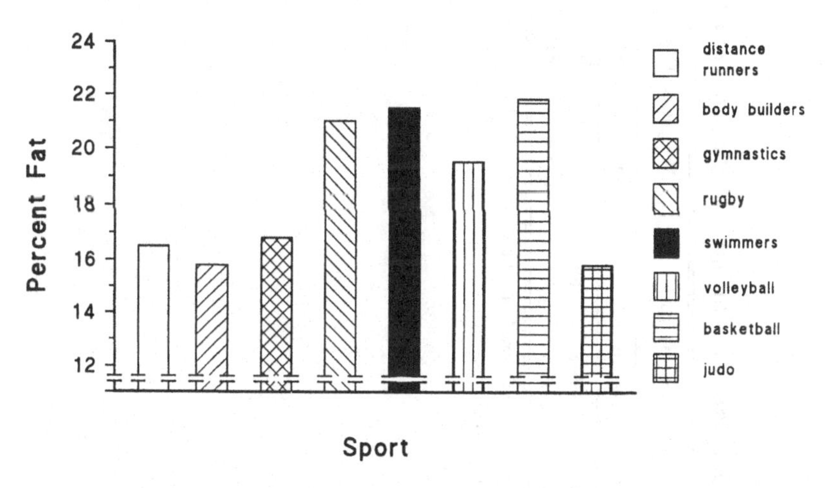

Fig. 10.4. Mean percentage fat in female athletes by sport.

man due to increased bone mass from heavy lifting in training, while the converse will be seen in the long distance runner. Hence, the percentage fat of the football player will be underestimated and the FFB will be overestimated; whereas the percentage fat of the long distance runner will be overestimated and the FFB will be underestimated.

Further, interpretation of gross FFB, without accounting for body size (i.e. height), is troublesome because a shorter person will probably have less FFB than a taller individual. This is particularly obvious in Fig. 10.2 where gymnasts are depicted as having less FFB than long distance runners.

Table 10.1. *Age and physical characteristics of the non-athletic males and females*[a]

Sample	Age (years)		Height (cm)		Weight (kg)		Fat-free body (kg)		% fat	
	Mean	SD	Mean	SD	Mean	SD	Mean	SD	Mean	SD
Males ($n = 116$)	24.2	3.4	178.8	6.8	76.5	9.2	65.4	7.8	14.4	5.8
Females ($n = 158$)	24.6	3.2	165.6	6.6	59.6	8.0	45.1	5.1	23.8	7.1

[a] The subjects used for the non-athletic population were from the University of Illinois, the University of Arizona, and the University of California at Berkeley.

Gymnasts are typically shorter than long distance runners, and relative to their height, gymnasts have much more FFB than long distance runners. An alternative method of characterising physique which controls for height is, therefore, recommended (Slaughter & Lohman, 1980; Slaughter *et al.*, 1987).

Relative musculoskeletal size

Since FFB is largely dependent upon height, a regression approach controlling for height effectively characterises musculoskeletal size and enables meaningful comparisons of physique between athletes and non-athletes, and among various athletic groups. The differences in physique can be expressed as deviations relative to the standard error of estimate (SEE) from a regression of FFB on height^x in a non-athletic sample. Positive deviations reflect more FFB for a given height, and negative deviations reflect less FFB for a given height.

Measures of body density and stature were obtained on college age males ($n = 116$) and females ($n = 158$) from three geographical regions in the United States: Arizona, California and Illinois. The age and descriptive characteristics of the non-athletic subjects are depicted in Table 10.1. Within gender, regressions of FFB on height were employed to determine the line of best fit for the male and female samples independently. A linear relationship was not assumed. The regression analyses yielded the following equations: $\hat{Y} = 0.00199$ (height^2) $+ 1.67$, SEE $= 6.2$; $\hat{Y} = 0.012709$ (height$^{1.6}$) $- 0.08$, SEE $= 4.2$ for the males and females, respectively (Figs. 10.5, 10.6). These results underscore the importance of using a regression approach rather than a simple ratio (FFB/height), as the assumptions underlying the use of ratios were violated. For example, the line of best fit for both samples was exponential rather than linear. Moreover, the slopes of the resulting regression lines are not similar to the slopes of the ratios (0.366 males, 0.272 females). Further, the intercepts of the regression equations are not zero.

Various college-age athletic groups were sequestered from the literature, and comparisons were made among the athletic groups in reference to the non-athletic regression of FFB on height. Only samples for which measures of body density and stature were reported were included in the analysis.

Deviations from the non-athletic regression of FFB on height in male American football players ranged from 4.6 kg (0.7 SEE) in college quarterbacks to 20.0 kg (3.2 SEE) in defensive linemen (Table 10.2, Fig. 10.7). It is interesting to note that the collegiate football players in 1984 have profiles characteristic of professional football players in 1972. The linemen are taller and have more FFB per unit of height than the other

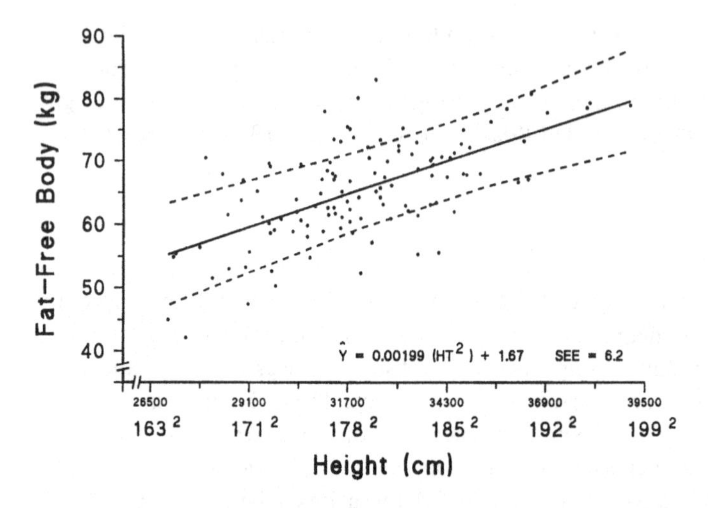

Fig. 10.5. Fat-free body to height relationship in non-athletic males (n = 116).

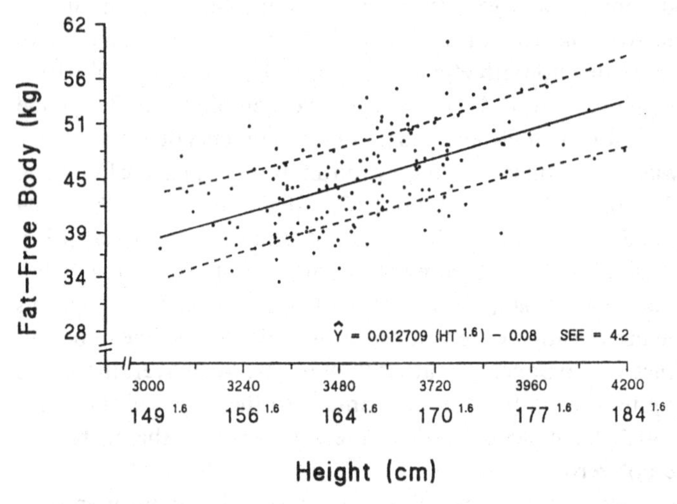

Fig. 10.6. Fat-free body to height relationship in non-athletic females (n = 158).

positions examined. Moreover, the offensive backs have more FFB relative to height than the defensive backs.

Differences in the relative musculoskeletal size of athletes in judo, cycling and ice hockey were detected, with cyclists deviating the least (− 0.6 kg, − 0.1 SEE) and judo participants deviating the most (15.1 kg, 2.4 SEE) from the non-athletic regression of FFB on height (Table 10.3, Fig. 10.8). Male body builders, power lifters and Olympic lifters had significantly

Table 10.2. *Deviations from non-athletic regression line of FFB on height in male American football players*

	n	Age (years)	Height (cm)	FFB (kg)	Predicted FFB[a] (kg)	FFB (kg) deviations from regression line	No. of SEEs (kg)
Professional[b]							
Defensive backs	26		182.5	76.5	67.9	8.6*	1.4
Offensive backs	40		183.8	81.9	68.9	13.0*	2.3
Linebackers	28		188.6	87.6	72.4	15.2*	2.5
Defensive line	32		193.0	95.8	75.8	20.0*	3.2
Offensive line	38		192.4	94.7	75.3	19.4*	3.1
Quarterbacks	16		185.0	77.1	69.8	7.3*	1.2
College[c]							
Defensive backs	10	20.0	183.9	76.8	69.0	7.8*	1.3
Offensive backs	14	20.8	182.1	82.1	67.7	14.4*	2.3
Linebackers	9	19.5	186.3	83.6	70.7	12.9*	2.1
Defensive line	11	19.5	193.0	95.6	75.8	19.8*	3.2
Offensive line	20	19.4	190.0	88.8	73.5	15.3*	2.5
Quarterbacks	4	19.5	186.1	75.2	70.6	4.6*	0.7

[a] Predicted FFB calculated from the equation $\hat{Y} = 0.00199\ (\text{ht}^2) + 1.67$; SEE = 6.2.
[b] From Wilmore & Haskell (1972).
[c] From Smith & Mansfield (1984).
* $p < 0.05$.

Table 10.3. *Deviations from the non-athletic regression line of FFB on height in male cycling, ice hockey and judo*

Investigator	n	Age (years)	Height (cm)	FFB (kg)	Predicted FFB[a] (kg)	FFB (kg) deviations from regression line	No. of SEEs (kg)
Judo							
Callister *et al.* (1990)	8	25.6	179.9	81.2	66.1	15.1**	2.4
Ice hockey							
Agre *et al.* (1988)							
Goalies	4	25.0	177.2	70.9	64.2	6.7**	1.1
Forwards	15	24.8	183.5	79.5	68.7	10.8**	1.7
Defense	8	24.9	184.7	77.7	69.6	8.1**	1.3
Cyclists							
Coyle *et al.* (1991)							
Elite	9	24.0	180.1	65.6	66.2	−0.6	−0.1
Good	6	23.0	176.3	65.4	63.5	1.9	0.3

[a] Predicted FFB calculated from the equation $\hat{Y} = 0.00199\ (\text{ht}^2) + 1.67$; SEE = 6.2.
$^{**}p < 0.01$.

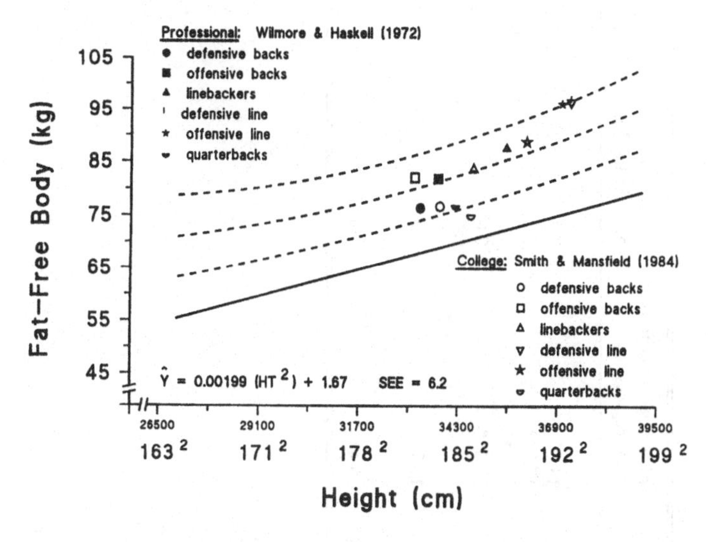

Fig. 10.7. Deviations from the non-athletic regression of fat-free body on height in male American football players.

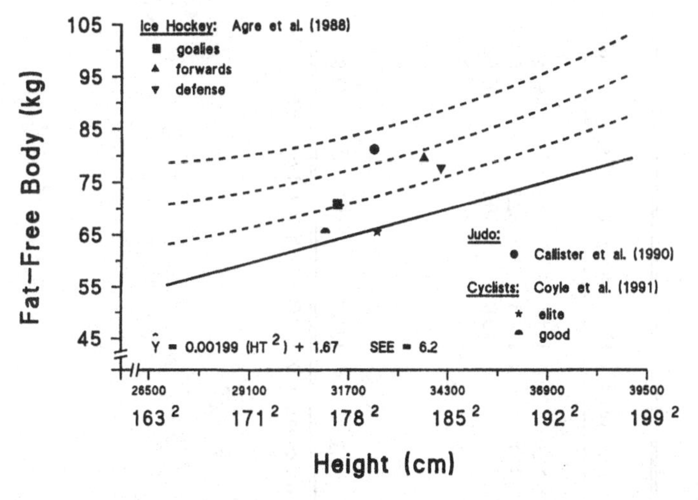

Fig. 10.8. Deviations from the non-athletic regression of fat-free body on height in male cyclists, ice hockey players, and judo participants.

Table 10.4. *Deviations from the non-athletic regression line of FFB on height in male body builders, power lifters and Olympic lifters*

Investigator	*n*	Age (years)	Height (cm)	FFB (kg)	Predicted FFB[a] (kg)	FFB (kg) deviations from regression line	No. of SEEs (kg)
Body builders							
Katch, *et al.* (1980)	18	27.8	177.1	74.6	62.6	14.0**	2.3
Spitler *et al.* (1980)	10	30.1	179.3	82.3	65.6	17.3**	2.7
Elliot *et al.* (1987)	16	25.0	175.0	70.3	62.6	7.7**	1.2
Sandoval *et al.* (1989)	5	25	176.7	76.3	63.8	12.5**	2.0
Power lifters							
Katch *et al.* (1980)	13	24.8	173.5	73.3	61.6	11.7**	1.9
Olympic lifters							
Katch *et al.* (1980)	8	25.3	173.9	68.2	61.8	6.4*	1.0

[a] Predicted FFB calculated from the equation $\hat{Y} = 0.00199\ (\mathrm{ht}^2) + 1.67$; SEE = 6.2.
*$p < 0.05$.
**$p < 0.01$.

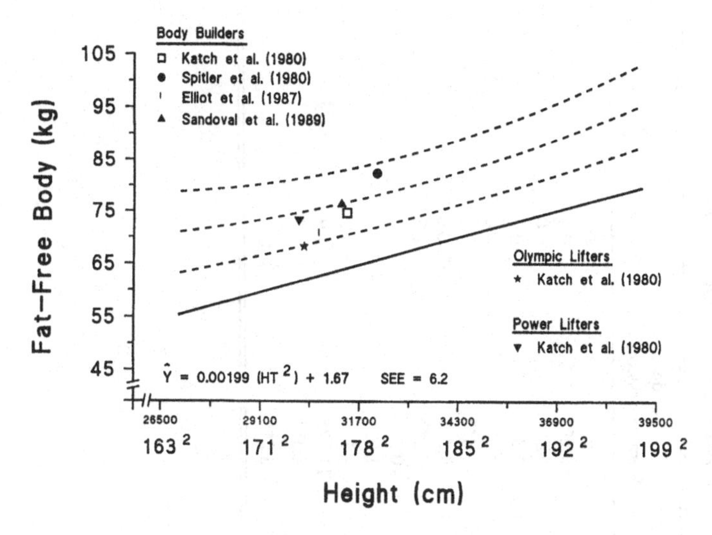

Fig. 10.9. Deviations from the non-athletic regression of fat-free body on height in male body builders, power lifters and Olympic lifters.

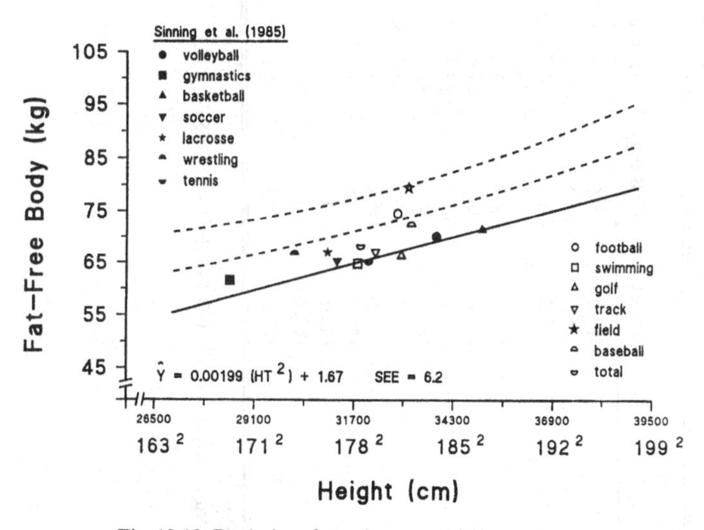

Fig. 10.10. Deviations from the non-athletic regression of fat-free body on height in college male athletic populations.

Table 10.5. *Deviations from non-athletic regression line of FFB on height in college male athletic populations*

	n	Height (cm)	FFB (kg)	Predicted FFB[a] (kg)	FFB (kg) deviations from regression line	No. of SEEs (kg)
Volleyball	15	184.0	70.0	69.0	1.0	0.2
Gymnastics	19	168.7	61.5	58.3	3.2**	0.5
Basketball	14	187.2	71.4	71.4	0.0	0.0
Soccer	18	176.8	65.2	63.9	1.3	0.2
Lacrosse	24	176.1	66.9	63.4	3.5**	0.6
Wrestling	44	173.6	66.4	61.6	4.8**	0.8
Tennis	9	179.1	65.3	65.5	−0.2	−0.03
American football	36	181.2	74.5	67.0	7.5**	1.2
Swimming	27	178.3	64.7	64.9	−0.2	−0.1
Golf	9	181.5	66.2	67.2	1.0	0.2
Track	20	179.6	67.0	65.9	1.1	0.2
Field	11	182.0	79.4	67.7	11.7**	1.9
Baseball	18	182.2	72.0	67.7	4.3**	0.7
Total	265	178.5	68.3	65.1	3.2**	0.6

Data from Sinning *et al.* (1985).

[a] Predicted FFB calculated from the equation $\hat{Y} = 0.00199\ (\text{ht}^2) + 1.67$; SEE = 6.2.

$^{**}p < 0.01$.

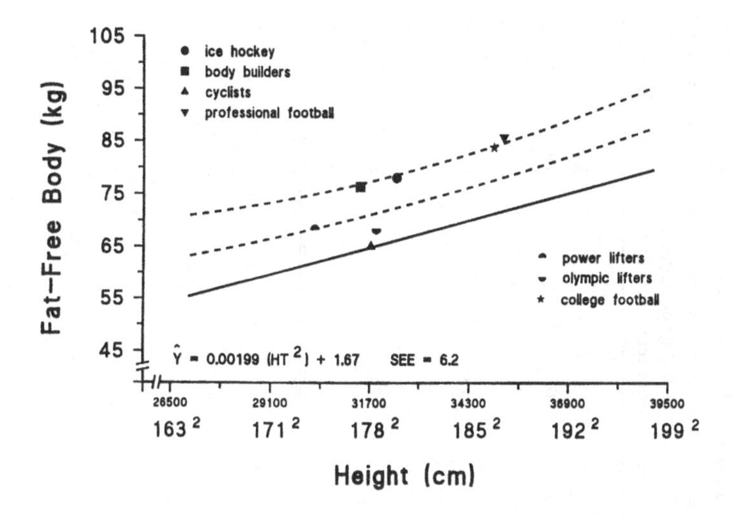

Fig. 10.11. Deviations from the non-athletic regression of fat-free body on height in male athletes.

more FFB relative to their height than the non-athletes. The deviations from the non-athletic regression of FFB on height ranged from 6.4 kg (1.0 SEE) in the Olympic lifters to 17.3 kg (2.7 SEE) in the body builders. Moreover, the lifters were typically shorter than the body builders (Table 10.4, Fig. 10.9).

Athletes who compete at a lower level generally have less FFB per unit of height than more élite competitors within the same sport. Deviations from the non-athletic regression of FFB on height in less élite male athletes ranged from − 0.2 kg (− 0.1 SEE) for the swimmers to 11.7 kg (1.9 SEE) for the field throwers (Table 10.5, Fig. 10.10).

The mean deviations from the male non-athletic regression of FFB on height of select athletic groups, pooling position within sport, are depicted in Fig. 10.11. American football players, followed by the ice hockey players and body builders, are taller and have more FFB relative to height than the other athletes considered. Power lifters are typically shorter, but have more FFB per unit of height, than the Olympic lifters. Finally, the FFB to height relationship observed in cyclists is similar to that observed in the non-athletic group.

The relative musculoskeletal size of select female athletic groups was also compared with that of the non-athletic females. Rugby players are generally shorter, but have more FFB per unit of height than the volleyball and basketball players (Table 10.6, Fig. 10.12). The deviations from the non-athletic regression of FFB on height ranged from 1.7 kg (0.4 SEE) for the volleyball players to 5.8 kg (1.4 SEE) for the basketball players.

Table 10.6. *Deviations from non-athletic regression line of FFB on height in female volleyball, basketball and rugby players*

Investigator	*n*	Age (years)	Height (cm)	FFB (kg)	Predicted FFB[a] (kg)	FFB (kg) deviations from regression line	No. of SEEs (kg)
Basketball							
Sinning (1973)	14	20.1	169.1	49.5	46.6	2.9	0.7
Walsh *et al.* (1984)	49	19.3	176.5	53.9	49.9	4.0*	1.0
Johnson *et al.* (1989)	8	20.1	177.7	55.7	50.5	5.2*	1.3
Rugby							
Sedlock *et al.* (1988)	17	20.6	165.1	50.6	44.9	5.8*	1.4
Volleyball							
Kovaleski *et al.* (1980)							
Varsity	10	20.5	171.0	49.1	47.4	1.7*	0.4
Junior Varsity	9	19.2	173.4	51.7	48.5	3.2*	0.8
Johnson *et al.* (1989)	14	20.1	175.2	54.5	49.3	5.2*	1.2

[a] Predicted FFB calculated from the equation $\hat{Y} = 0.0127\ (ht^{1.6}) - 0.08$; SEE = 4.2.
*$p < 0.05$.

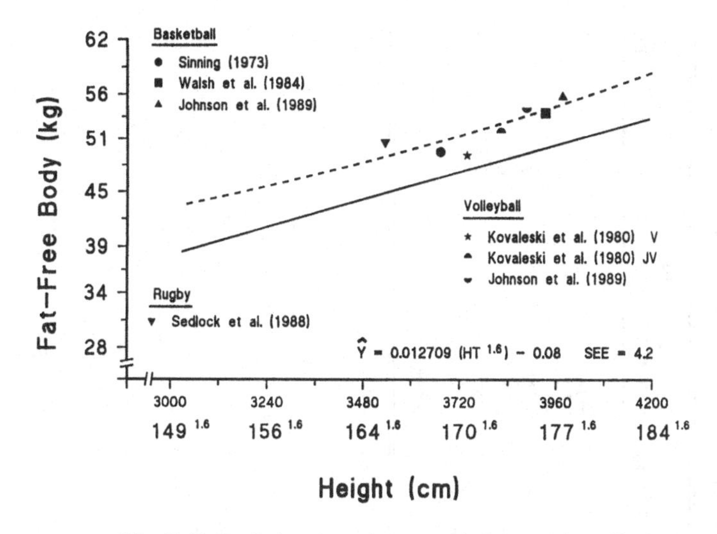

Fig. 10.12. Deviations from the non-athletic regression of fat-free body on height in female volleyball, basketball and rugby players.

Interestingly, the 1989 women volleyball and basketball players are taller and have greater relative musculoskeletal size than their predecessors in 1973 and 1980.

The deviations from the non-athletic regression of FFB on height ranged from 2.7 kg (0.7 SEE) to 5.5 kg (1.3 SEE) in female body builders, whereas a greater deviation was observed in the rugby players (9.0 kg, 2.1 SEE) (Table 10.7, Fig. 10.13). The deviations from the non-athletic regression of FFB on height in gymnasts were within 1 SEE and ranged from − 0.5 kg (− 0.1 SEE) to 3.9 kg (0.9 SEE) (Table 10.8, Fig. 10.14). Interestingly, the relative musculoskeletal size of a recent United States collegiate team (Rector, 1991) is similar to that of an earlier national team from former Czechoslovakia (Sprynarova & Parizkova, 1969).

In contrast, the deviations from the non-athletic regression of FFB on height in female long distance runners were all negative, and ranged from − 3.4 kg (− 0.8 SEE) to − 0.4 kg (− 0.1 SEE) (Table 10.9, Fig. 10.15). Hence, female long distance runners have less FFB per unit of height than their non-athletic counterparts. The deviations from the non-athletic regression of FFB on height in female swimmers ranged from 2.0 kg (0.5 SEE) to 6.3 (1.5 SEE), whereas a negative deviation (− 3.0 kg, − 0.7 SEE) was observed in the synchronised swimmers (Table 10.10, Fig. 10.16).

Hence, when the female athletic groups were considered, the greatest FFB per unit of height was observed in the power lifters, whereas the lowest relative musculoskeletal size was detected in the long distance runners

Table 10.7. *Deviations from the non-athletic regression line of FFB on height in female body builders and power lifters*

Investigator	n	Age (years)	Height (cm)	FFB (kg)	Predicted FFB[a] (kg)	FFB (kg) deviations from regression line	No. of SEEs (kg)
Body builders							
Freedson *et al.* (1983)	10	27	160.8	46.2	43.0	3.2*	0.8
Elliot *et al.* (1987)	16	28	164.0	47.1	44.4	2.7*	0.7
Sandoval *et al.* (1989)	6	27	162.9	48.1	43.9	4.2**	1.0
Johnson *et al.* (1990)	10	30.4	165.2	48.1	44.9	3.2**	0.8
Heinrich *et al.* (1990)	10	25.7	167.1	51.2	45.7	5.5**	1.3
Power lifters							
Johnson *et al.* (1990)	10	25.2	164.6	53.6	44.6	9.0**	2.1

[a] Predicted FFB calculated from the equation $\hat{Y} = 0.0127\,(\mathrm{ht}^{1.6}) - 0.8$; SEE = 4.2.
*$p < 0.05$.
**$p < 0.01$.

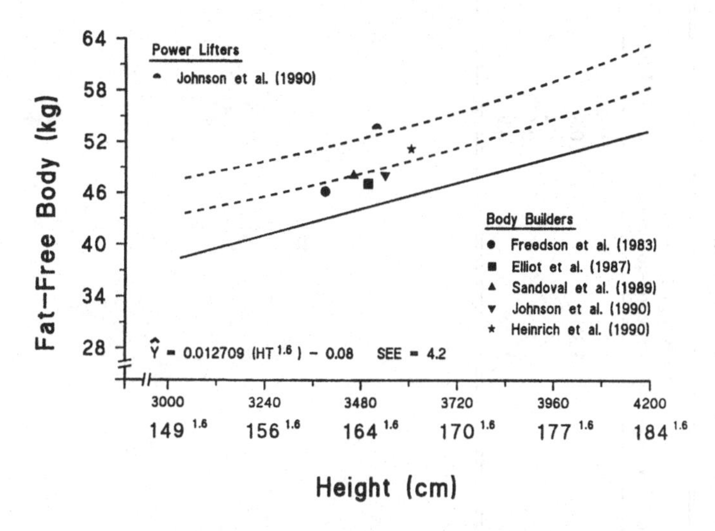

Fig. 10.13. Deviations from the non-athletic regression of fat-free body on height in female body builders and power lifters.

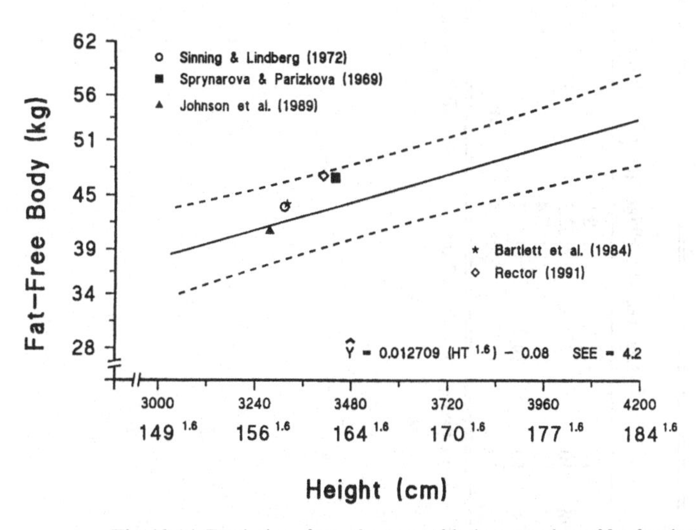

Fig. 10.14. Deviations from the non-athletic regression of fat-free body on height in female gymnasts.

Table 10.8. *Deviations from the non-athletic regression line of FFB on height in female collegiate gymnasts*

Investigator	*n*	Age (years)	Height (cm)	FFB (kg)	Predicted FFB[a] (kg)	FFB (kg) deviations from regression line	No. of SEEs (kg)
Sinning & Lindberg (1972)	14	20.1	158.5	43.7	42.0	1.7	0.4
Sprynarova & Parizkova (1969)	10		162.3	46.9	43.6	3.3*	0.8
Johnson *et al.* (1989)	8		157.4	41.0	41.5	−0.5	−0.1
Bartlett *et al.* (1984)	10	19.7	158.7	44.0	42.1	1.9	0.5
Rector (1991)	13	19.7	161.4	47.1	43.2	3.9**	0.9

[a] Predicted FFB calculated from the equation $\hat{Y} = 0.0127\ (ht^{1.6}) - 0.08$; SEE = 4.2.

*$p < 0.05$.

**$p < 0.01$.

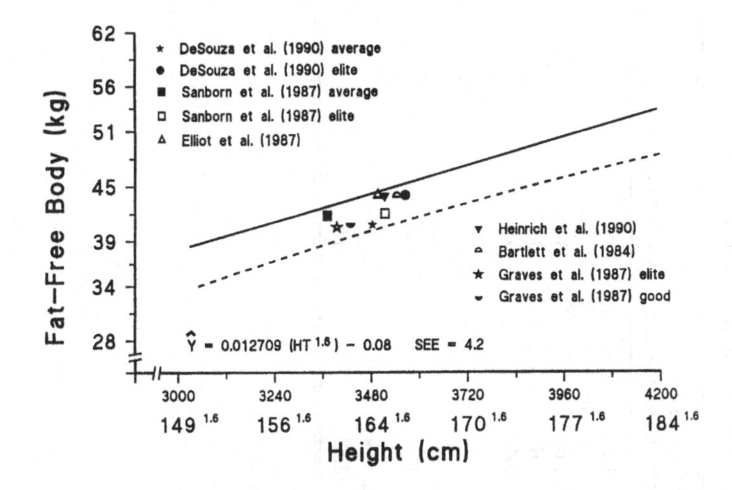

Fig. 10.15. Deviations from the non-athletic regression of fat-free body on height in female long distance runners.

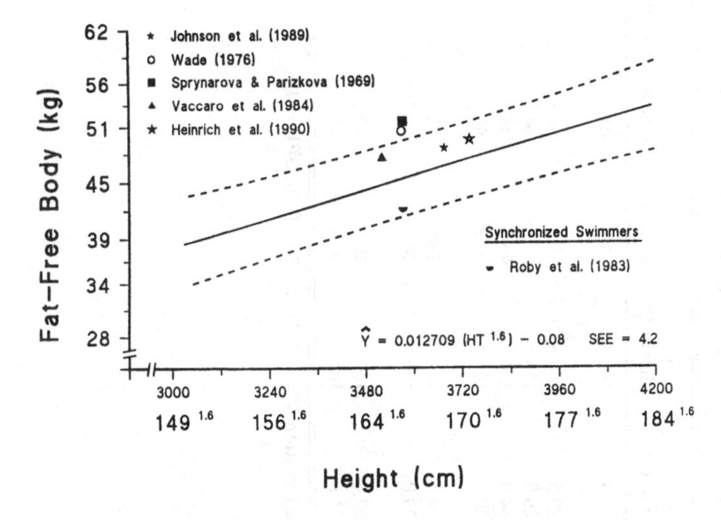

Fig. 10.16. Deviations from the non-athletic regression of fat-free body on height in female swimmers and synchronised swimmers.

Table 10.9. *Deviations from the non-athletic regression line of FFB on height in female long distance runners*

Investigator	n	Age (years)	Height (cm)	FFB (kg)	Predicted FFB[a] (kg)	FFB (kg) deviations from regression line	No. of SEEs (kg)
Bartlett *et al.* (1984)	10	20.3	165.4	43.8	45.0	−1.2	−0.3
Sanborn *et al.* (1987)							
Amenorrhoeic	7	29.5	160.3	41.8	42.8	−1.0	−0.2
Eumenorrhoeic	7	26.9	164.5	42.0	44.6	−2.6	−0.6
Elliot *et al.* (1987)	14	34.0	164.0	44.0	44.4	−0.4	−0.1
Heinrich *et al.* (1990)	5	20.2	164.5	43.9	44.6	−0.7	−0.2
De Souza *et al.* (1990)							
Amenorrhoeic	8	24.5	163.6	40.8	44.2	−3.4	−0.8
Eumenorrhoeic	8	29.0	166.0	44.0	45.2	−1.2	−0.3
Graves *et al.* (1987)							
Elite runners	15	27.6	161.0	40.5	43.2	−3.2*	−0.8
Good runners	12	25.6	162.0	41.0	43.5	−2.5*	−0.6

[a] Predicted FFB calculated from the equation $\hat{Y} = 0.0127\ (ht^{1.6}) - 0.08$; SEE = 4.2.
*$p < 0.05$.

Table 10.10. *Deviations from the non-athletic regression line of FFB on height in female collegiate swimmers and synchronised swimmers*

Investigator	n	Age (years)	Height (cm)	FFB (kg)	Predicted FFB[a] (kg)	FFB (kg) deviations from regression line	No. of SEEs (kg)
Swimmers							
Wade (1976)	15	20.2	166.1	50.5	45.3	5.2**	1.2
Sprynarova & Parizkova (1969)	10		166.2	51.6	45.3	6.3**	1.5
Vaccaro *et al.* (1984)	10	20.0	164.7	47.5	44.7	2.8*	0.7
Johnson *et al.* (1989)	7	19.3	169.2	48.6	46.6	2.0	0.5
Heinrich *et al.* (1990)	13	21.7	171.0	49.5	47.4	2.1	0.5
Synchronised swimmers							
Roby *et al.* (1983)	13	20.1	166.2	42.3	45.3	−3.0	−0.7

[a] Predicted FFB calculated from the equation $\hat{Y} = 0.0127\ (ht^{1.6})\ -0.08$; SEE = 4.2.
$^{*}p < 0.05$.
$^{**}p < 0.01$.

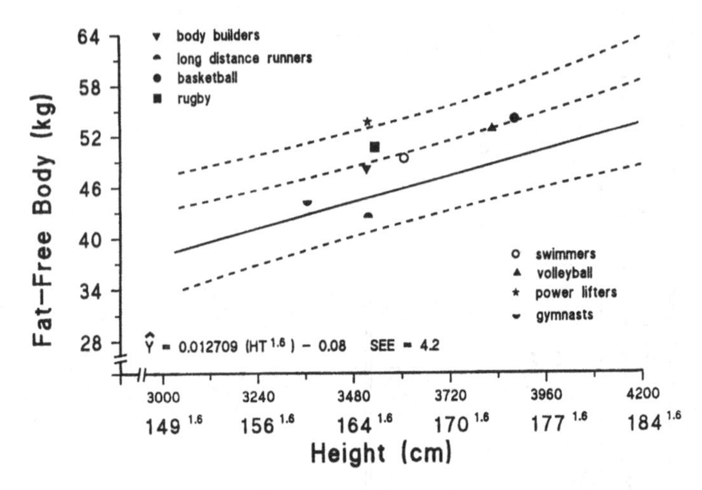

Fig. 10.17. Deviations from the non-athletic regression of fat-free body on height in male athletes.

(Fig. 10.17). Note that when height was controlled in the expression of FFB, it is clear that the gymnasts have a greater relative musculoskeletal size than the long distance runners (Fig. 10.17), even though they have less total FFB (Fig. 10.2).

Summary

The three independent, yet interrelated components of physique, namely, body size, structure and composition, were discussed with regard to their measurement, their importance, and their impact on physical performance. Implicit in the use of ratios to depict body structure are three basic assumptions which are not upheld empirically. Moreover, gross body composition measures (e.g. FFB) which do not control for body size (e.g. height) can be misleading. Since FFB is largely dependent upon height, a regression approach which controls for height is recommended to characterise musculoskeletal size as it enables meaningful comparisons of physique between athletes and non-athletes, and among different athletic groups. The difference in physique can be expressed as deviations relative to the SEE from a regression of FFB on height. Of the athletic groups considered, FFB per unit of height was greatest in the American football players and rugby players for the males and females, respectively; whereas the cyclists and long distance runners had the least amount of FFB per unit of height in the males and females, respectively. Physique appears to be dependent upon the particular sport, level of competition and gender considered.

Precise estimates of body physique are useful in determining physical readiness for sport participation, establishing profiles of specific athletic groups, and for ascertaining optimal weight for optimal performance and health. Although physical training can enhance performance, genetic endowment must not be overlooked. It is unlikely that a sumo wrestler can transform his physique sufficiently to become a successful marathon runner.

References

Agre, J.C., Casal, D.C., Leon, A.S., McNally, M.C., Baxter, T.L. & Serfass, R.C. (1988). Professional ice hockey players: Physiologic, anthropometric, and musculoskeletal characteristics. *Archives of Physical Medicine and Rehabilitation*, **69**, 188–92.

Asmussen, E. (1974). Developmental patterns in physical performance capacity. In: *Fitness, Health and Work Capacity*, ed. L.A. Larsen, pp. 435–48. New York: Macmillan.

Asmussen, E. & Heeboll-Nielsen, R.R. (1955). A dimensional analysis of physical performance and growth in boys. *Journal of Applied Physiology*, **7**, 593–603.

Bartlett, H.L., Mance, M.J. & Buskirk, E.R. (1984). Body composition and expiratory reserve volume in female gymnasts and runners. *Medicine and Science in Sports and Exercise*, **16**, 311–15.

Benn, R.T. (1971). Some mathematical properties of weight-for-height indices used as measures of adiposity. *British Journal of Preventative and Societal Medicine*, **25**, 42–50.

Boileau, R.A. & Lohman, T.G. (1977). The measurement of human physique and its effect on phyiscal performance. *Orthopedic Clinics of North America*, **8**, 563–81.

Boileau, R.A., Lohman, T.G., Slaughter, M.H., Ball, T.E., Going, S.B. & Hendrix, M.K. (1984). Hydration of the fat-free body in children during maturation. *Human Biology*, **56**, 651–666.

Boileau, R.A., Lohman, T.G. & Slaughter, M.H. (1985). Exercise and body composition of children and youth. *Scandanavian Journal of Sports Sciences*, **7**, 17–27.

Boileau, R.A., Slaughter, M.H., Stillman, R.J. *et al.* (1992). Age-related variability in the fat-free body water and mineral content of 20–70 year old adults. *FASEB Journal*, **6**, A1673 (abstract).

Bosaeus, I. & Isaksson, B. (1991). Body composition methodology: dilutometry. In: *Nutritional Status Assessment: a Manual for Population Studies*, ed. F. Fidanza, pp. 71–83. London: Chapman and Hall.

Brodie, D.A. (1988*a*). Techniques of measurement of body composition, part I. *Sports Medicine*, **5**, 11–40.

Brodie, D.A. (1988*b*). Techniques of measurement of body composition, part II. *Sports Medicine*, **5**, 74–98.

Bunt, J.C., Going, S.B., Lohman, T.G., Heinrich, C.H., Perry, C.D. & Pamenter, R.W. (1990). Variation in bone mineral content and estimated body fat in young adult females. *Medicine and Science in Sports and Exercise*, **22**, 564–9.

Callister, R., Callister, R.J., Fleck, S.J. & Dudley, G.A. (1990). Physiological and

performance responses to overtraining in elite judo athletes. *Medicine and Science in Sports and Exercise*, **22**, 816–24.
Carter, J.E.L. & Heath, B.H. (1990). *Somatotyping: Development and Applications.* Cambridge: Cambridge University Press.
Coyle, E.F., Feltner, M.E., Kautz, S.A. *et al.* (1991). Physiological and biomechanical factors associated with elite endurance cycling performance. *Medicine and Science in Sports and Exercise*, **23**, 93–107.
De Souza, M.J. Maguire, M.S., Rubin, K.R. & Maresh, C.M. (1990). Effects of menstrual phase and amenorrhea on exercise performance in runners. *Medicine and Science in Sports and Exercise*, **22**, 575–80.
Elliot, D.L., Goldberg, L., Kuehl, K.S. & Catlin, D.H. (1987). Characteristics of anabolic-androgenic steroid-free competitive male and female bodybuilders. *The Physician and Sports Medicine*, **15**, 169–79.
Fidanza, F., ed. (1991). *Nutritional Status Assessment: a Manual for Population Studies*. London: Chapman & Hall.
Fomon, S.J., Haschke, F., Ziegler, E.E. & Nelson, S.E. (1982). Body composition of reference children from birth to age 10 years. *American Journal of Clinical Nutrition*, **35**, 1169–75.
Forbes, G.B. (1987). *Human Body Composition: Growth, Aging, Nutrition and Activity*. New York: Springer-Verlag.
Freedson, P.S., Mihevic, P.M., Loucks, A.B. & Girandola, R.N. (1983). Physique, body composition, and psychological characteristics of competitive female body builders. *The Physician and Sports Medicine*, **11**, 85–93.
Garn, S.M., Leonard, W.R. & Hawthorne, V.M. (1986). Three limitations of the body mass index. *American Journal of Clinical Nutrition*, **44**, 119–21.
Garrow, J.S. & Webster, J. (1985). Quetelet's index (W/H^2) as a measure of fatness. *International Journal of Obesity*, **9**, 147–53.
Graves, J.E., Pollok, M.L. & Sparling, P.B. (1987). Body composition of elite female distance runners. *International Journal of Sports Medicine*, **8**, 96–102.
Gunther, B. (1975). Dimensional analysis and theory of biological similarity. *Physiological Reviews*, **55**, 659–74.
Haschke, F., Fomon, S.J. & Ziegler, E.E. (1981). Body composition of a nine-year-old reference boy. *Pediatric Research Journal*, **15**, 847–9.
Heath, B.H. & Carter, J.E.L. (1967). A modified somatotype method. *American Journal of Physical Anthropology*, **27**, 57–74.
Heinrich, C.H., Going, S.B., Pamenter, R.W., Perry, C.D., Boyden, T.W. & Lohman, T.G. (1990). Bone mineral content of cyclically menstruating female resistance and endurance trained athletes. *Medicine and Science in Sports and Exercise*, **22**, 558–63.
Heymsfield, S.B. (1991). Body composition methodology: advanced methods. In: *Nutritional Status Assessment: a Manual for Population Studies*, ed. F. Fidanza, pp. 83–96. London: Chapman & Hall.
Johnson, G.O., Nebelsick-Gullett, L.J., Thorland, W.G. & Housh, T.J. (1989). The effect of a competitive season on the body composition of university athletes. *Journal of Sports Medicine and Physical Fitness*, **29**, 314–19.
Johnson, G.O., Housh, T.J., Powell, D.R. & Ansorge, C.J. (1990). A physiological comparison of female body builders and power lifters. *Journal of Sports*

Medicine and Physical Fitness, **30**, 361–4.
Katch, V.L., Katch, F.I., Moffatt, R. & Gittleson, M. (1980). Muscular development and lean body weight in body builders and weight lifters. *Medicine and Science in Sports and Exercise*, **12**, 340–44.
Keys, A., Fidanza, F., Karvonen, M.J. Kimura N. & Taylor, H.L. (1972). Indices of relative weight and obesity. *Journal of Chronic Disease*, **25**, 329–43.
Kovaleski, J.E., Parr, R.B., Hornak, J.E. & Roitman, J.L. (1980). Athletic profile of women college volleyball players. *The Physician and Sports Medicine*, **8**, 112–18.
Lohman, T.G. (1986). Applicability of body composition techniques and constants for children and youth. In: *Exercise and Sports Sciences Reviews*, vol. 14, ed. K.B. Pandolf, pp. 325–357. New York: Macmillan.
Lohman, T.G. (1992). *Advances in Body Composition Assessment*. Champaign, Illinois: Human Kinetics.
Lohman, T.G., Boileau, R.A. & Slaughter, M.H. (1984*a*). Body composition in children and youth. In: *Advances in Pediatric Sport Sciences*, vol. 1, ed. R.A. Boileau, pp. 29–57. Champaign, Illinois: Human Kinetics.
Lohman, T.G., Slaughter, M.H., Boileau, R.A., Bunt, J.C. & Lussier, L. (1984*b*). Bone mineral measurements and their relation to body density in children, youth and adults. *Human Biology*, **56**, 667–79.
Norgan, N.G., ed. (1985). *Human Body Composition and Fat Distribution*. Report of an EC Workshop, London 10–12 December 1985, pp. 64–70. Wageningen: Stichting Nederlands Instituut voor de Voeding.
Norgan, N.G. (1991). Anthropometric assessment of body fat and fatness. In *Anthropometric Assessment of Nutritional Status*, ed. J.H. Himes, pp. 197–212. New York: A.R. Liss.
Parnell, R.W. (1958). *Behaviour and Physique*. London: Edward Arnold.
Rector, A. (1991). Differences in physique between collegiate female gymnasts and nongymnasts. Unpublished master's thesis, University of Illinois, Urbana, Illinois.
Roby, F.B., Buono, M.J., Constable, S.H., Lowdon, B.J. & Tsao, W.Y. (1983). Physiological characteristics of champion synchronized swimmers. *The Physician and Sports Medicine*, **11**, 136–147.
Sanborn, C.F., Albrecht, B.H. & Wagner, W.W. (1987). Athletic amenorrhea: lack of association with body fat. *Medicine and Science in Sports and Exercise*, **19**, 207–12.
Sandoval, W.M., Heyward, V.H., & Lyons, T.M. (1989). Comparison of body composition, exercise and nutritinoal profiles of female and male body builders at competition. *Journal of Sports Medicine and Physical Fitness*, **29**, 63–70.
Sedlock, D.A., Fitzgerald, P.I. & Knowlton, R.G. (1988). Body composition and performance characteristics of collegiate women rugby players. *Research Quarterly for Exercise and Sport*, **59**, 78–82.
Seidell, J.C. (1991). Environmental influences on regional fat distribution. *International Journal of Obesity*, **15**(S2), 31–5.
Sheldon, W.H. (1940). *The Varieties of Human Physique*. New York: Harper and Brothers.
Sheldon, W.H., Dupertuis, C.W. & McDermott, E. (1954). *Atlas of Men*. New York: Harper and Brothers.
Sinning, W.E. (1973). Body composition, respiratory function, and rule changes in

women's basketball. *The Research Quarterly*, **44**, 313–21.

Sinning, W.E. & Lindberg, G.D. (1972). Physical characteristics of college age women gymnasts. *The Research Quarterly*, **43**, 226–34.

Sinning, W.E., Dolny, D.G., Little, K.D. *et al.* (1985). Validity of 'generalized' equations for body composition analysis in male athletes. *Medicine and Science in Sports and Exercise*, **17**, 124–30.

Slaughter, M.H. & Lohman, T.G. (1980). An objective method for measurement of musculo-skeletal size to characterize body physique with application to the athletic population. *Medicine and Science in Sports and Exercise*, **12**, 170–4.

Slaughter, M.H., Lohman, T.G., Christ, C.B. & Boileau, R.A. (1987). An objective method for the measurement of musculo-skeletal size in children and youth. *Journal of Sports Medicine and Physical Fitness*, **27**, 461–72.

Slaughter, M.H., Lohman, T.G., Boileau, R.A., Christ, C.B. & Stillman, R.J. (1990). Differences in the subcomponents of the fat-free body in relation to height between black and white children. *American Journal of Human Biology*, **2**, 209–17.

Smith, J.E. & Mansfield, E.K. (1984). Body composition prediction in university football players. *Medicine and Science in Sports and Exercise*, **16**, 398–405.

Spitler, D.L., Diaz, F.J., Horvath, S.M. & Wright, J.E. (1980). Body composition and maximal aerobic capacity of body builders. *Journal of Sports Medicine and Physical Fitness*, **20**, 181–8.

Sprynarova, S. & Parizkova, J. (1969). Comparison of the functional, circulatory and respiratory capacity in girl gymnasts and swimmers. *Journal of Sports Medicine and Physical Fitness*, **9**, 165–72.

Thompson, D.W. (1917). *On Growth and Form*. Cambridge: Cambridge University Press.

Vaccaro, P., Ostrove, S.M., Vandervelden, L., Goldfarb, A.H. & Clarke, D.H. (1984). Body composition and physiological responses of master female swimmers 20 to 70 years of age. *Research Quarterly for Exercise and Sport*, **55**, 278–84.

Van Itallie, T.B., Yang, M.U., Heymsfield, S.B., Funk, R.C. & Boileau, R.A. (1990). Height-normalized indices of the body's fat-free mass and fat mass: potentially useful indicators of nutritional status. *American Journal of Clinical Nutrition*, **52**, 953–9.

Wade, C.E. (1976). Effects of a season's training on the body composition of female college swimmers. *Research Quarterly*, **47**, 292–5.

Walsh, F.K., Heyward, V.H. & Schau, C.G. (1984). Estimation of body composition of female intercollegiate basketball players. *The Physician and Sports Medicine*, **12**, 74–86.

Wilmore, J.H. & Haskell, W.L. (1972). Body composition and endurance capacity of professional football players. *Journal of Applied Physiology*, **33**, 564–7.

11 *The assessment of the body composition of populations*

N.G. NORGAN

Introduction

The measurement of body composition occurs in many branches of biology and medicine. It is measured by the human biologist studying human variation and adaptation and it is being used increasingly in the assessment of nutritional and growth status, fitness, work capacity, disease and its treatment. In human energetics, it is widely used for the standardisation of variables such as basal metabolic rate or physical work capacity in, for example, the investigation of the types and scope of adaptation to chronic energy and nutrient deficiency or excess.

Most determinations of body composition, whether for research or surveillance purposes, are made in the field as opposed to the laboratory or bedside. There is then a clear need for simple, rapid, safe, accurate methods to determine body composition in population studies. For many years anthropometry has, to some extent, met this need. Body composition has properties additional to those of size, i.e. weight and linear dimensions, but weight/height indices have been and continue to be used widely as proxies for body composition measurements. The best known of these is the Body mass index (BMI) or Quetelet's index.

Skinfold thicknesses stand in their own right as indices of fat and fatness and, by difference, of leanness. There are reference data for all ages and both sexes. The practice of estimating body composition from anthropometric variables is fraught with pitfalls as many of the published equations have not been proved to be of general validity. The effects of age and ethnic group on the relationships between skinfolds and total body fatness are not well established and there is little information on whether estimation procedures drawn up on Europeans can be applied to the rest of the world. These are considered to be some of the most important areas for investigations in the 1990s (Lohman, 1992).

Anthropometry is being challenged by a number of 'black box' devices that appear simple and rapid and which produce a result without the need for understanding its calculation and origin. Experiences with these vary and they have not been widely used on populations other than Europeans.

In this chapter, the techniques available for assessment of body composition of large groups of individuals (more than 50 per day) in the clinic or in the field are considered. Particular attention is given to the question of the characteristics of good estimation procedures and how they are identified. There has been very little work on the applicability of body composition techniques for population assessment to the various ethnic groups, so a detailed examination has been made of the performance in other groups of estimation procedures drawn up on Europeans. The use of skinfolds and BMI are considered in detail, followed by the techniques of bioelectrical impedance analysis, near-infrared interactance and ultrasound. Their accuracy and why they may perform unsatisfactorily are discussed. Recommendations on choice of methods for children, adults and the elderly are included.

Methods of measuring body composition

Our concepts and descriptions of body composition have followed the methods available for measurement. This has led to descriptions of the body in terms of chemical or anatomical components, in terms of the whole body or regions, and in terms of two-compartment models. Wang *et al.* (1992) have made an important step forward in clarifying thinking and dispelling misconceptions with their description of body composition as a five-level model: the atomic, molecular, cellular, tissue-system and whole-body level. They regard body composition research as a branch of human biology with three interconnecting areas: measurement techniques, the biological factors that influence body composition and the as yet inadequately formulated area of body composition levels and their organisational rules. The model not only clarifies the concept of body composition but also reveals gaps in knowledge leading to the identification of new research areas.

For population assessment, body composition methods can be considered as either whole-body or regional approaches. These are listed in Table 11.1. Of the *in vivo* whole-body techniques listed, only bioelectrical impedance analysis (BIA) can be considered suitable for population assessment, i.e. simple and portable. In contrast, most of the regional techniques, e.g. anthropometry such as skinfold thicknesses (SKF) and circumferences (C), plus near-infrared interactance (NIRI) and ultrasound (US) are, on the same basis, suitable. However, the interest is usually in whole-body composition and to provide this information all regional techniques require transformation, as illustrated towards the bottom of Table 11.1. Much of the error of these estimations is introduced at this stage, but there are recognised approaches to transformation that improve the chances of the technique being of acceptable accuracy. (The term 'estimation' is used in preference to 'prediction' as it is a reminder that there is always some

Table 11.1. *Methods of measuring of body composition*

1. *In vitro methods*
   Anatomical dissection
   Chemical analysis
2. *In vivo methods*
   Whole body
   Density
   Body water
   Potassium
   Elemental, (calcium, nitrogen, carbon)
   Electrical properties: bioelectrical impedance analysis (BIA), total body electrical conductivity (TOBEC)
   Absorptiometry: dual photon (DPA), dual energy radiography (DER)
   Magnetic resonance imaging
   Regional techniques
   Depths and thicknesses, e.g. subcutaneous adipose tissue: skinfold calipers (SKF), ultrasound (US), near-infrared interactance (NIRI)
   Cross-sectional areas: skinfolds + circumferences, computed tomography (CT), B-scan ultrasound
   Volumes and masses: skinfolds + circumferences + lengths, displacement, segmental BIA, CT, DER, 3D body scanning

   Indirect estimations
   Transformation of anthropometry or regional estimates to give whole body estimate, e.g. body mass index (BMI) → body composition

   Skinfolds → density → fat mass + fat-free mass
   Body density $= a + b$ ($\log_{10} \sum$ four skinfolds)
   (e.g. Durnin & Womersley, 1974)

   Body density $= a - b1\ (\sum \text{SKF}) + b2\ (\sum \text{SKF})^2 - b3\ (\text{age})$
   (e.g. Jackson & Pollock, 1978)

   BIA → body water

   NIRI → % fat

imprecision.) If these approaches are followed in the development of the technique and can be identified then the selection of suitable methods is facilitated.

Choosing a method for estimating body composition

To be able to choose an appropriate body composition method it is necessary to know how estimation techniques should be established, how those techniques that are being considered were established and what is known about their performance in populations or individuals similar to those to be investigated. The procedure for establishing a method of estimating body composition is to measure body composition by a criterion method, such as body density, as the dependent variable, followed by the estimation of independent variable(s), such as anthropometry, in a biologically homogeneous group. The requirements for producing equations generalisable to other populations have been discussed (Katch & Katch, 1980; Lohman, 1981; Norgan & Ferro-Luzzi, 1985) and reiterated (Norgan, 1991; Lohman, 1992). It is now recognised that the best equations are obtained when large numbers of subjects (> 50) are used and when the independent variables are selected using either a hierarchical, or an all possible subset ridge or robust regression procedure as opposed to stepwise regression or author selection, in order to avoid problems of multi-colinearity.

Similarly, a better-informed choice of method is possible where:

1. The standard error of estimate ($SEE = SDy \cdot \sqrt{(1 - r^2)}$ was reported. This is more informative than the correlation coefficient, r, which is influenced by the range of the data.
2. The derived equations were validated internally on a separate subsample.
3. Reliability studies have been conducted on each method estimating the sources of variation due to trial, time of day, exercise, diet, menstrual cycle, etc.
4. More than one criterion method or a multicomponent method was used.
5. Interlaboratory studies have been conducted on the same subjects with standardised methodology.

No body composition technique has been put through such a testing procedure before being used on other populations. Some have been tested extensively by 'case history' with the passage of time: for example, the equations of Durnin & Womersley (1974), Jackson & Pollock (1978) and Jackson *et al.* (1980). However, validation studies have become much better designed and more comprehensive. Notable examples are those of Sinning *et al.* (1985) and Pierson *et al.* (1991) on skinfolds and those of Lukaski *et al.* (1986) and the interlaboratory study of Segal *et al.* (1988) on BIA.

The performance of an equation can be assessed by internal or external cross-validation as suggested by Lohman (1981, 1992). The answers sought are:

1. Are the means of the measured and estimated scores significantly different?
2. Are the SD of measured and estimated scores comparable? Estimated scores tend to be overestimated at the bottom of the

Table 11.2. *Validation statistics for the estimation of percentage fat (% fat) and fat-free mass (FFM) (mean ± SD) of Papua New Guinea men and women using the equation of Durnin & Womersley (1974)*

	Men (n = 120)		Women (n = 88)	
	% fat	FFM (kg)	% fat	FFM (kg)
Measured	10.0 ± 3.5	51.5 ± 5.1	21.5 ± 4.3	38.5 ± 4.5
Estimated	11.7 ± 2.6	50.4 ± 4.5	22.0 ± 3.7	38.3 ± 3.9
SEE	3.4	2.2	3.6	1.9
TE	4.1	2.4	3.9	2.0
t	5.24 ($p<0.001$)	5.3 ($p<0.001$)	1.10 (NS)	1.13 (NS)
r	0.26	0.90	0.54	0.91

SD, standard deviation; SEE, standard error of the estimate; TE, total error; t, t statistic; r, correlation coeficient.

range and underestimated at the top, leading to a restricted range which is identified by a low SD of the estimated scores

3. Do SEE and r compare favourably with the original statistics and with those of other studies and equations?
4. What is the total or root mean square error (RMSE) = $\sqrt{(\sum(Y_{est} - Y_{meas})^2/N)}$? This combines the difference of the means and the SEE.

An example of these validation statistics for the estimation of percentage of the body weight as fat (% fat) and fat-free mass (FFM) in Papua New Guineans (Norgan *et al.*, 1982), using the equations of Durnin & Womersley (1974), is shown in Table 11.2. Measured and estimated % fat and FFM are both significantly different in the men ($p<0.001$) but not the women. The SD of estimated scores is less than that of the measured, and for % fat the correlation coefficients are low. Using the subjective rating scale of Lohman (1992) to assess the accuracy of anthropometric estimations in young men and women, the SEE for % fat are 'good' to 'fairly good' and those of FFM are rated as 'excellent' or 'ideal'. This is a rather strange mixture and the scale, shown in Table 11.3, may prove to require some amendment.

Another approach to assessing estimation procedures is the graphical display of estimated and measured scores and the calculation of some simple statistics. Bland & Altman (1986) proposed plotting the difference between estimated and measured scores against their mean. The mean is used as it is regarded as the best estimate of the true score, which is unknown. This is debatable but it avoids a common statistical artefact arising from the difference being related to the individual scores, and the

Table 11.3. *Lohman's subjective ratings of the accuracy of estimations of fat-free mass (FFM) and % fat*

SEE of FFM (kg)		SEE of % fat	
Men	Women	Men and women	Rating
2.0–2.5	1.5–1.8	2.0	Ideal
2.5	1.8	2.5	Excellent
3.0	2.3	3.0	Very good
3.5	2.8	3.5	Good
4.0	3.2	4.0	Fairly good
4.5	3.6	4.5	Fair
>4.5	>4.0	5.0	Not recommended

Adapted from 'Anthropometric assessment of fat-free body mass' by T.G. Lohman. In *Anthropometric Assessment of Nutritional Status* (p. 175), ed. J.H. Himes, 1991. New York: Wiley (1991, reprinted by permission of John Wiley & Sons, Inc.) and *Advances in Body Composition Assessment: Current Issues in Exercise Science*, Monograph Number 3 (p. 4) by T.G. Lohman, 1992. Champaign, Ill: Human Kinetics Publishers (Copyright 1992 Timothy G. Lohman, reprinted with permission.)

approach has proved popular. The plot can show the extent of the mean difference and whether error depends on the level of the body composition variable. In particular, it emphasises the often wide limits of agreement of estimated and measured scores of individuals, in contrast to group means. The data for % fat of the Papua New Guinea women is shown in Fig. 11.1. There is a bias (mean difference) of + 0.5 % fat, as seen in Table 11.2. The 95% confidence interval of the bias is – 0.4 to – 1.3% fat and the limit of agreement (2 × SD of the differences) is 7.8% fat, so that one or two values are more than 8% from the average of the two methods. The correlation coefficient is low (0.18), indicating that differences are unrelated to the level of fatness.

A final method of testing estimation procedures is to use analysis of covariance to compare the regression coefficient and intercept of regression equations relating regional anthropometry to whole-body composition to identify differences between groups.

The estimation of the body composition of non-Europeans

There has been little work on developing procedures for population assessment of body composition in non-European groups, even though they comprise the majority of the world's population. Consequently, the question arises as to whether estimation procedures developed on Europeans can be applied to other groups. Few studies have investigated this topic but

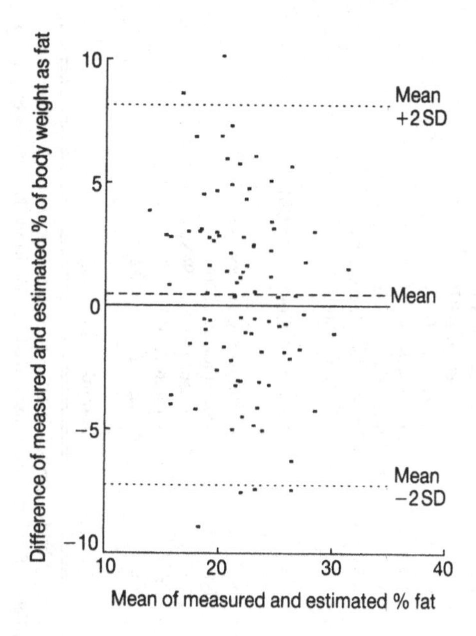

Fig. 11.1. Plot of the differences in estimated and measured % fat in Papua New Guinea women against the mean of the two scores, showing the mean difference and ± 2 SD of the mean.

there are data in the literature that allow comment to be made. However, different criterion methods and variables have been used by different authors and the estimation procedures have involved height, weight, circumferences and skinfolds, or BIA and NIRI. As each comparison may be affected by the age, sex or fatness of the individual or population and their health and fitness, it requires a voluminous literature before the question can be answered satisfactorily.

Skinfold equations

Table 11.4 shows the results of the use of Durnin & Womersley's equations drawn up on European men and women for estimating the body composition of other groups. The body composition method is listed, as is the body composition component compared and the result of estimation, the effect. An upward-pointing arrow signifies on overestimation of the component, a downward-pointing arrow an underestimation. Where known, the difference in measured and estimated means given in component units, i.e. 2% fat refers to 2% fat units, not 2% difference in the body fat means. Estimations are regarded as accurate where the estimated and predicted means are not

Table 11.4. *Do Durnin & Womersley's equations estimate accurately the body composition of groups other than Europeans?*

	n	Method	Comparison	Effects	Accurate?	Authors
Men						
N. American Blacks	250	Density	% fat	+ 2.5% (sign ?)	No ?	Friedl & Vogel (1991)
N. American Blacks	45	D_2O	% fat	+0.7 %	Yes	Zillikens & Conway (1990)
N. American Blacks	45	D_2O	% fat	Using a Lange caliper, + 1.8%	No	Zillikens & Conway (1990)
Gambians	20	D_2O	FFM	+ 1.2 kg	Yes	Minghelli *et al.* (1990)
New Guineans	73	Density	% fat	+ 1.1%	Yes	Norgan *et al.* (1982)
New Guineans	49	Density	% fat	+ 3.9%	No	Norgan *et al.* (1982)
Guatemalans	124	Density	% fat	Wide limits of agreement,↑	No	Immink *et al.* (1992)
Colombians	11	D_2O	% fat	↓	No ?	Spurr *et al.* (1981)
Chileans	31	Density	% fat	− 0.3 %	Yes	Apud & Jones (1980)
S. Indians	30	Density	Regression	↓, NS	Yes	Jones *et al.* (1976)
Uttar Pradeshi	30	Density	Regression	↓, NS	Yes	Jones *et al.* (1976)
Rajputs	30	Density	Regression	↓, NS	Yes	Jones *et al.* (1976)
Gurkhas	30	Density	Regression	↓	No	Jones *et al.* (1976)
Chinese	47	Density	% fat	+ 2.0%	No	Eston, R.G. (personal communication)
Women						
N. American Blacks	45	D_2O	% fat	0.1%	Yes	Zillikens & Conway (1990)
N. American Blacks	45	D_2O	% fat	Using a Lange caliper, + 1.7%	No	Zillikens & Conway (1990)
Beninese	24	D_2O	Fat mass	+ 1.5 kg	No	Schultinck *et al.* (1992)
Gambians	60	D_2O	FFM	+ 0.2 kg	Yes	Lawrence *et al.* (1988)
New Guineans	68	Density	% fat	+ 0.2%	Yes	Norgan *et al.* (1982)
New Guineans	20	Density	% fat	+ 1.3%	Yes	Norgan *et al.* (1982)
Guatemalans	24	Density	% fat	Wide limits of agreement, ↑	No	Immink *et al.* (1992)

D_2O, hydrometry using deuterated water; ↑ or +, mean estimated greater than mean measured body composition; ↓ or −, mean estimated less than mean measured body composition; Yes, mean estimated body composition not significantly different from the mean measured composition; No, mean estimated body composition significantly different from the mean measured composition.

significantly different, the limits of agreement are narrow or the regressions are not significantly different.

Friedl & Vogel (1991) did not test their data so the significance is not established (hence the question mark) and Spurr *et al.* (1981) reported no difference but the published means and standard deviations suggest otherwise. For the Gambians of Minghelli *et al.* (1990) the anthropometry was performed 8 months after the hydrometry but the authors discount the effects that seasonality might have had on the findings. The two groups of New Guineans represent individuals under and over 30 years of age. Jones *et al.* (1976) used analysis of covariance of the regressions of body density and the log sum of four skinfolds of Durnin & Womersley (1974). The lines were not significantly different except in the Gurkhas, a Himalayan rather than an Indian group. The effect seems to be about + 2% fat.

The wide limits of agreement for the Guatemalans of Immink *et al.* (1992) are a slightly different approach in that emphasis is placed on individual results, in this case those with the greatest errors, as opposed to mean results. RMSE of 3.9 and 4.9% fat and 2.3 and 2.5 kg FFM for men and women respectively can be compared with the ratings of Lohman (1992) in Table 11.3, bearing in mind that RMSE are different, usually higher, statistics to SEE. The estimation of % fat in women is 'not recommended' but is 'fairly good' for men. For FFM, the estimation is 'good' for women and 'ideal' for men. This echoes the findings in Table 11.2 of low accuracy for fat but high accuracy for FFM and calls into question the rating scale.

What comes out of this review of published studies is that roughly half find the Durnin & Womersley (1974) equations satisfactory and half do not. The studies in Table 11.4 are arranged according to ethnic group but there seems to be no effect of ethnicity, or of group size, criterion method or recency (in the expectation that more recent studies would be better designed) on whether the estimation is successful. This type of 'vote counting' to review or sum up research has a number of pitfalls. Equal weight is given to each study although studies vary considerably in numbers of subjects, 'rigour', etc. A further difficulty is the problem of publication bias – that studies showing 'no difference' are less likely to appear in the literature. Also, work in theses is not included. The area of 'summing-up' is well described by Light & Pillemer (1984).

Table 11.5 shows the experiences from using other equations and approaches. Here there is more consistency in the finding that the equations are not applicable to ethnic groups other than those on which they were derived, although there are contradictory findings. Hortobagyi *et al.* (1992) found the seven skinfold equation of Jackson & Pollock (1978) estimated accurately the mean % fat of North American Black male athletes but Friedl & Vogel (1991) found a significant difference in Black service men.

Table 11.5. *Do other equations drawn up on Europeans estimate accurately the body composition of groups other than Europeans?*

Group	*n*	Method	Equation[a]	Comparison	Effects	Accurate?	Authors
Men							
N. American Blacks	55	Density	8	% fat	0.0%	Yes	Hortobagyi *et al.* (1992)
N. American Blacks	250	Density	1	Density	−1%	Yes?	Friedl & Vogel (1991)
N. American Blacks	250	Density	Mean of 2–7	% fat	−2%	No ?	Friedl & Vogel (1991)
N. American Blacks	250	Density	8	% fat	−4%	No ?	Friedl & Vogel (1991)
N. American Blacks	45	D_2O	8	% fat	−2.9%	No	Zillikens & Conway (1990)
N. American Blacks	140	Density	Authors'	Regressions	↓D	No	Vickery *et al.* (1988)
N. American Blacks	15	Density	Mean of 2–7	Density	−0.0112 kg/l	No	Schutte *et al.* (1984)
Colombians	11	D_2O	4	% fat	↓	No	Spurr *et al.* (1981)
Inuit (Eskimos)	33	D_2O		low *r* with skinfolds	–	No	Shephard *et al.* (1973)
S. Indians	30	Density	4, 6, 7	% fat	Each ↓	No	Jones *et al.* (1982)
Rajputs	30	Density	4, 6, 7	% fat	Each ↓	No	Jones *et al.* (1982)
Gurkhas	30	Density	4, 6, 7	% fat	Each ↓	No	Jones *et al.* (1982)
Indians (16M + 9F)	25	Density	9	high *r*		Yes	Sen & Banerjee (1958)
Taiwanese	28	Density		Regression		Yes	Parizkova (1977)
Women							
N. American Blacks	98	Density	10	% fat	+ 0.7%	Yes	Sparling *et al.* (1993)
N. American Blacks	45	D_2O	10	% fat	−2.9%	No	Zillikens & Conway (1990)
Inuit (Eskimos)	41	D_2O		low *r* with skinfolds	–	No	Shephard *et al.* (1973)
Indians	65	Density	12 equations	Density	↓	(No)	Satwanti *et al.* (1977)

[a] Equations: 1, Jackson & Pollock (1978) using 7 skinfold thicknesses and 2 circumferences; 2, Katch & McArdle (1973); 3, Lohman (1981); 4, Pascale *et al.* (1956); 5, Pollock *et al.* (1976); 6, Sloan (1967); 7, Wilmore & Behnke (1969); 8, Jackson & Pollock (1978) equation with 7 skinfolds; 9, Brozek & Keys (1951); 10, Jackson *et al.* (1980).

The decision for Satwanti *et al.* (1977) is not clear-cut as they report very high lung residual volumes during the determination of body density.

Why the equations may not work

There are a number of reasons why skinfold estimation equations may not perform satisfactorily in any group. These can be considered as methodological (both technical and statistical) and biological. Whereas it may be possible to reduce the influence of the former, the latter are more difficult to allow for. Technical factors include variations in technique, e.g. in measurement of skinfolds. Choice of calipers may be important, Lange calipers giving higher mean % fat than Harpenden (Lohman *et al.*, 1984) or Holtain (Zillikens & Conway, 1990) calipers. Statistical factors include using small sample size, large numbers of independent variables, untransformed data and lack of cross-validation. These do no apply to the equations of Durnin & Womersley (1974) and Jackson and colleagues (Jackson & Pollock, 1978; Jackson *et al.*, 1980). However, no estimation equations have been drawn up on randomly selected subjects.

The important biological factors in the case of skinfold equations are likely to be the fat distribution (the proportion of fat situated subcutaneously), fat patterning (the distribution of subcutaneous adipose tissue over the surface) and skinfold compressibility, all of which are affected by age, sex and ethnic group. The effects of some of these are illustrated in Fig. 11.2. Mean individual skinfolds from the papers of Immink *et al.* (1992) and Ulijaszek *et al.* (1989) have been used to calculate % fat from the equations of Durnin & Womersley (1974) using a single sum at the different ages for each of the different combinations of skinfold thicknesses. For Durnin & Womersley's subjects, whether two, three or four skinfolds were used did not affect the value of the % fat estimate at any age. In the Guatemalan men and Wopkaimin women of Papua New Guinea, two skinfolds yields % fat values different from those obtained using three or four skinfolds, due to differences in fat patterning. The effect in the two groups is, however, opposite: in Guatemalan men they are higher, in Papua New Guinean women they are lower.

In many human biological investigations the ages in the population may not be known. There is then a tendency to use the total age group equations of Durnin & Womersley (1974), but this should be resisted. Fig. 11.2 shows that for the same sum of skinfolds different % fat values are obtained for the different age groups. There are a variety of reasons for this, the principal one being the internal–external redistribution of fat in adult life. As most of the subjects of Durnin & Womersley were young, the overall age group equations produce % fat values close to those of young subjects. If the

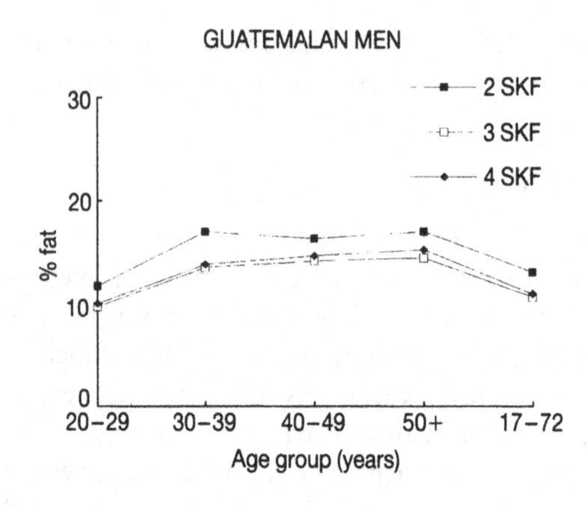

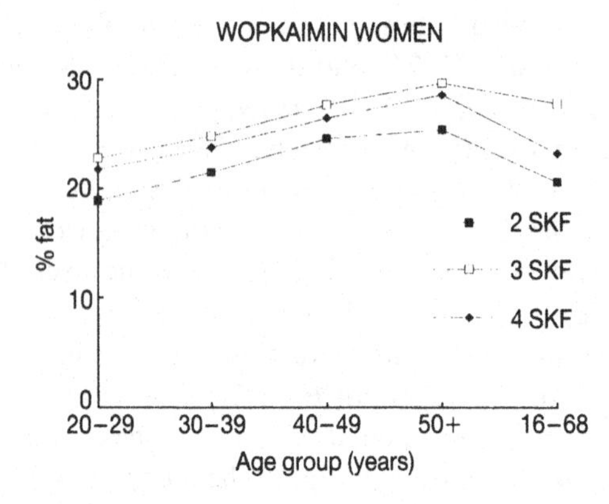

Fig. 11.2. The effects of age and fat patterning on the estimation of % fat by the equations of Durnin & Womersley (1974) in Guatemalan men (Immink *et al.*, 1992) and Wopkaimin (Papua New Guinea) women (Ulijaszek *et al.*, 1989). SKF, skinfolds.

subjects are of unknown age but known to be middle aged there would be less error using the 40–50 years than the 17–72 years equations. It is noticeable that the % fat for 16- to 68-year-old women from log sum of three skinfolds appears higher than might be expected. This can be observed in the data of Durnin & Womersley too, which suggests there may be an error in the equation.

One point often overlooked in estimating or comparing estimates of body composition in groups other than those on which the estimation equations are based is that the SEE describes the variation about the whole of the regression line but the error of an estimate (usually referred to as the standard error of the forecast) increases as the measurement moves away from the midpoint. Thus, the error in the thin and fat, in athletes and the obese, is higher than for those in the middle of the range, such that specific equations may be required for these groups. In particular, the estimation of body composition in lean populations in developing countries may be poorer because of this effect. Also, a restricted range of body composition in any subgroup will result in lower correlations between measured and estimated variables, even if there are no other biological differences.

It is also important to remember that the criterion methods are not completely accurate. It has been calculated that 25–50% of the error in estimation may arise from the criterion method (Lohman, 1992), arising from the variability of FFM composition, amongst other things.

Body mass index

Dissatisfaction with relative weight and weight/height ratios as indices of overweight and obesity was an important stimulus to the development of body composition methods. However, weight/height ratios such as the body mass index (BMI, kg/m^2) are widely used as indices of overweight and fatness, particularly in epidemiological studies. Recently, limits of BMI indicative of chronic energy deficiency have been proposed (James *et al.*, 1988; Ferro-Luzzi *et al.*, 1992). The relationship between BMI and body composition is weaker than, for example, body composition and skinfolds – although not always so, particularly in women and the elderly. The nature of the relationship is far from straightforward as BMI is correlated with measures of both leanness and of fatness and it is as much a measure of size as it is of composition. Some variation in BMI is caused by differences in shape, such as relative leg length, further complicating its interpretation (Norgan, 1994*a*,*b*) and additional anthropometric measures may be necessary to improve its interpretation.

Summaries of the data on BMI (and other weight/height indices) and body composition in Europeans have been assembled (Willett, 1990; Shetty & James, 1994). The data on Italian shipyard workers is an illustration (Norgan & Ferro-Luzzi, 1982). These show that BMI is well correlated with fatness (% fat; $r = 0.75$) but more so with fat content (fat mass; $r = 0.87$) and, to a significant degree, leanness (fat-free mass; $r = 0.68$). Adding age increases significantly, by 13%, the proportion of the variance explained. In these Italian men, aged 22–55 years, the correlation of BMI and % fat equalled that of the sum of four skinfold thicknesses with % fat.

Are there ethnic differences in the relationships of BMI to body composition?

A fundamental question is whether different population groups have similar BMI–body composition relationships, i.e. the same % fat at a given BMI. A modelling approach suggested that large and important differences existed in the relationships of BMI to body composition in Ethiopians, Indians and Papua New Guineans (Norgan, 1990). However, the marked curvilinearity of the theoretical relationships were in contrast to the linear relations observed in populations. A number of reasons were suggested for this and, indeed, if an assumption is made of variable composition of differences in individuals and populations, the model yields linear solutions to the relationships (Norgan, 1994*c*).

The relationships of BMI and body composition in different ethnic groups have been reviewed recently (Norgan, 1994*c*). There were insufficient data on measured body composition in different ethnic groups so this was estimated using the Durnin & Womersley (1974) skinfold equations. In many cases BMI and sum of skinfold thicknesses were calculated from group means, although the weaknesses of this approach must be borne in mind. Data on 173 samples of men and 112 of women in mainly non-urban communities in developing countries were classified as (1) African, (2) Asian or 'Asian origin', which included native Americans, (3) Indo-Mediterraneans but not Europeans, and (4) Pacific peoples. Taken as a group, the two sexes showed relationships similar to those of Europeans. The regression statistics, correlations and SEE between BMI and body composition are shown in Table 11.6. According to the subjective rating of Lohman (1992), the estimation of % fat and FFM would be rated as 'fairly good'.

Within the Indo-Mediterraneans and Pacific groupings, similar correlations and SEE values to those in Europeans were observed, but in the African grouping the range of BMI was narrower and the correlations reduced. The range was also restricted in the Asian and 'Asian origin' grouping, but there was considerable scatter. Considering that a large proportion of the world's population would fall into this grouping, there is a need for many more data on Asians. The relationships of BMI and body composition in the four groups were tested for difference by analysis of covariance. Significant differences in intercept were found, the biological significance of which is not clear. There was no difference in the regression coefficients of % fat on BMI in the four groups of women and of FFM on BMI in men. In all other cases, differences were significant.

There are several reasons why different relationships may be found in different populations. One of these that has received little attention is

Table 11.6. *The relationships between body mass index (kg/m²) and body composition in* 176 *samples of men and* 112 *samples of women of mainly rural origin from developing countries*

	Men				Women			
	b	*c*	*r*	SEE	*b*	*c*	*r*	SEE
% fat	1.45	− 17.8	0.68	4.1 %	1.58	− 8.6	0.82	4.2%
Fat mass	1.44	− 22.8	0.78	3.1 kg	1.76	− 24.4	0.92	2.8 kg
Fat-free mass	1.22	− 24.3	0.64	3.9 kg	0.9	− 18.0	0.76	3.0 kg

b, regression coeficient; *c*, intercept; *r*, correlation coeficient; SEE, standard error of the estimate.

differences in body shape (Norgan, 1994*a*, *b*). It can be shown using a modelling approach that varying leg length or sitting height (SH) causes variations in stature and in shape, as indicated by the sitting height:stature ratio (SH/S); the range of variation in SH/S that is found in different populations is associated with differences in BMI of some 5 kg/m^2, equivalent to some 7.5% fat. Africans are relatively longer-legged and have lower SH/S than, for example, Asians, in which case BMI would indicate lower levels of fatness in Africans than actually existed. However, within ethnic groups there is as much variation, if not more, than between groups, so that a simple correction is unjustified except in extreme groups or individuals.

In conclusion, BMI and weight:height indices in general are the vectors of several effects. They have attributes of size and shape in addition to those of composition. They are as much measures of fat content and lean content as of fatness. Not surprisingly, there are ethnic differences in the relationships between BMI and body composition which may be a reflection of the individual differences that exist in all populations.

Bioelectrical impedance analysis

Bioelectrical impedance analysis (BIA) meets many of the requirements for a population method of estimating body composition, being quick, safe, portable, acceptable to the subject and inexpensive. It is described fully in the review by Baumgartner *et al.* (1990) and in the chapter by Deurenberg in this volume. The history of its introduction and the testing of the method illustrate the improvements that have occurred in methodology of body composition, stimulated by the seminal papers of Katch & Katch (1980) and Lohman (1981). However, debate continues as to whether or how much extra variation is explained by BIA in addition to that explained by simple anthropometric measures such as height and weight. At present,

Table 11.7. *Does bioelectrical impedance analysis (BIA) estimate accurately the body composition of groups other than Europeans?*

Group	n	Method	Comparison	Effect	Accurate?	Authors
Men						
N. American Blacks, athletes	55	Density	% fat	+ 5.4%	No	Hortobagyi *et al.* (1992)
N. American Blacks	43	D_2O	Own White equation	TBW, −1.0 l	No	Zillikens & Conway (1991)
Guatemalan highlanders	49	Density		low r, high RMSE	No	Diaz *et al.* (1991)
Guatemalans, as above	49	Density	Adding H^2/R to regression, NS	None	No	Diaz *et al.* (1989)
Women						
N. American Blacks	98	D_2O	% fat	+ 6.8%	No	Sparling *et al.* (1993)
N. American Blacks	45	D_2O	Own White equation	TBW, −0.7 l	No	Zillikens & Conway (1991)
Beninese	24	D_2O	Fat mass	+ 1.7 kg	No	Schultink *et al.* (1992)
Guatemalan highlanders	49	Density		Low r, high RMSE	No	Diaz *et al.* (1991)
Guatemalans, including 49 as above	99	Density	Adding H^2/R to regression, NS	None	No	Diaz *et al.* (1989)

D_2O, hydrometry using deuterated water; RMSE, root mean square error; TBW, total-body water.

BIA would seem to lie somewhere between BMI and skinfold thicknesses in terms of accuracy, its exact position depending on how the measurements are made and transformed and on whom they are made. Much more care regarding the conditions and application of the method is required than was first thought or might be thought. It is important to follow the practices used in deriving the estimation equation, the same type of machine, the same side of the body, the same limb position relative to the trunk and the same electrode placement. The ambient temperature and the level of hydration of the subject must be controlled. The estimation equations built into the apparatus by the manufacturers are generally regarded as inaccurate and population specific.

BIA has not been subject to validation in many non-European populations. The results are shown in Table 11.7. Apart from studies on North American Blacks, the studies are limited to Guatemalans and one group of West Africans. However, there is a consistency of finding: that BIA yields measures of body composition significantly different from those of density and hydrometry, by overestimating fat content and underestimating lean. All these studies used the RJL apparatus and the manufacturer's conversion equation, except where indicated. Zillikens & Conway (1991) also tested the equations of Kushner & Schoeller (1986), Lukaski *et al.* (1986) and Van Loan & Mayclin (1987), but without success. Interestingly, Schultinck *et al.* (1992) found bias between hydrometry and anthropometry and hydrometry and BIA but not between anthropometry and BIA. It could be that it is the criterion method, in this case hydrometry, that is not applicable to the Beninese rather than the estimation procedures.

The possible reasons for the poor showing of BIA in groups other than Europeans are easy to list but difficult to identify. The method assumes that the body is a conductor of homogeneous composition, fixed cross-sectional area and uniform current density distribution (Baumgartner *et al.*, 1990). The human body does not have these characteristics and the estimation equations have been drawn up empirically. However, significant deviations in tissue resistivity or shape could limit the accuracy of the estimation. In the case of North American Blacks, SH/S ratio is lower than in Whites, there is a higher density of FFM, higher FFM/height ratio and higher skin resistance. These represent systematic deviations from the theoretical and empirical whole-body conductor upon which the equations are based.

Near-infrared interactance

Near-infrared interactance (NIRI) has seen less trial than BIA. It has been regarded as satisfactory by some (Brodie & Eston, 1993) but not others (Israel *et al.*, 1989; Elia *et al.*, 1990; Fuller *et al.*, 1992; Hortobagyi *et al.*,

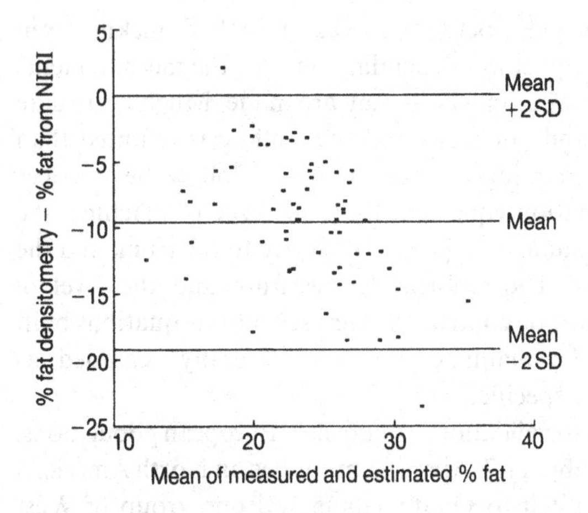

Fig. 11.3. The estimation of % fat of 63 middle-aged European men by near-infrared interactance (NIRI) compared with densitometry. From Brooke-Wavell (1992).

1992; Heyward *et al.*, 1992), with significant underestimation of fat content the common finding. Here, too, the contribution of information extra to that from height and weight is in question. In our experience in 63 middle-aged men, the Futurex-5000 apparatus underestimated % fat by 9.6% ($p<0.001$), with increasing error with higher % fat (Brooke-Wavell, 1992). The data are shown in Fig. 11.3. The recent use of multi-site NIRI is not proving an improvement on the original method of measuring interactance at the biceps site (Hortobagyi *et al.*, 1992).

Ultrasound

Ultrasound has a place in body composition studies when values of uncompressed subcutaneous adipose tissue thickness or information on its structure are required, but for population assessment of body composition it is no more informative than other population methods and it is more difficult to perform. The presence of fascial planes in subcutaneous adipose tissue (Alexander & Dugdale, 1992) renders the interpretation of the echoes difficult other than to the experienced observer. The method has been reviewed recently (Ramirez, 1992).

Visual assessment

Most body composition investigators have a degree of confidence in their ability to estimate the body composition of subjects by looking at them.

Three studies have quantified this ability by asking experienced observers to rate photographs. In those that used two raters (Sterner & Burke, 1986; Eckerson *et al.*, 1992) one rater performed as well or better than skinfold thicknesses, while the other did not. In the study of Hogdon *et al.* (1989), SEE values of body composition from the mean ratings of 11 raters were 4.3% and 4.0% in 860 men and 209 women, respectively. These fall somewhere between skinfolds and BMI in accuracy. The subjective assessment of body composition of populations is worthy of further investigation and may prove to have a place in some types of studies.

Assessment of body composition in children and the elderly

The assessment of the body composition of children and the elderly requires investigation and attention, but it is not intended to treat the subject systematically or extensively here. This need arises less from neglect, although this has certainly occurred, than from the operation of many biological factors during infancy, childhood and old age. Considerable changes occur over these periods which affect the applicability of criterion methods of body composition assessment and the nature of the relationship between anthropometry and body composition. Methods for adults applied to children result in overestimation of % fat and underestimation of FFM. The methods available for infants and children have been reviewed (Boileau *et al.*, 1985; Slaughter *et al.*, 1988; Davies & Preece, 1988; Ross Conference, 1989). A cross-validation study of laboratory-based criterion methods by Nielsen *et al.* (1993) found the most appropriate to be the four-component model of Boileau *et al.* (1985) or, if test facilities are limited, the two-component model of Lohman (1989).

The Wageningen group have produced estimation equations for 2- to 20-year-olds based on the Durnin & Womersley skinfold thicknesses that have statistics as good as those of adults (Weststrate & Deurenberg, 1989; Deurenberg *et al.*, 1990*b*). In particular, the equations allow for the changing composition and density of FFM in childhood. The equations of Slaughter *et al.* (1988) do this too and have been cross-validated.

Deurenberg and colleagues have also reported estimation equations using BIA for children of the same ages (Deurenberg *et al.*, 1990*a*). Houtkooper *et al.* (1989) concluded that BIA together with anthropometry was reliable and of acceptable accuracy for estimating FFM and % fat in 10- to 14-year-old children. In very young children it may be difficult to achieve the 4–5 cm separation necessary to prevent electrode polarisation and artificially high resistances.

Estimation of body composition from BMI has been proposed for 17- to 72-year-olds by Womersley & Durnin (1977) and for 7- to 83-year-olds by

Deurenberg *et al.* (1991). The estimation error in children and the elderly has been found to be comparable to that in young and middle-aged adults (Deurenberg *et al.*, 1989, 1991; Houtkooper *et al.* 1989). Usually, the relationships between body composition and anthropometry appear to be less strong in the elderly than in young adults.

A critical appraisal of criterion methods for elderly subjects and the effect the choice of criterion method has on estimation equations have been made by Heymsfield and colleagues (Heymsfield *et al.*, 1989: Baumgartner *et al.*, 1991). They consider that the new technologies allow multi-component measures of body composition that overcome the problems of changing FFM composition with age, and that estimation equations for the elderly should be based on such criterion methods. The assessment of body composition of those over 60 years old by BIA also requires age-specific equations (Deurenberg *et al.*, (1990*c*).

Recommendations

There are a number of factors to consider in choosing a method for measuring large numbers of people quickly and accurately in the laboratory or in the field, over and above those considered earlier in the section on 'Choosing a method for estimating body composition'. Some of these are listed in Table 11.8. The methods available and the feasibility (ease of use) and desirability (information gained) of each are shown in Table 11.9. Desirable methods tend to be less feasible under most circumstances, but anthropometric techniques score highly on both scales.

When to use anthropometry and when to transform

Human biologists have been recommended to use anthropometry itself, rather than attempt to make estimates of whole-body composition from equations, as even if equations provide useful estimates for mean parameters for samples they are not reliable for individual estimation (Johnston, 1982). Estimation involves multiplying anthropometric measurements by a series of constants with the aim of producing the variables we want or with which we are most comfortable. These constants may vary between studies for both biological and technical reasons. The need for a measure of whole-body composition is usually less real than we imagine. For many purposes (e.g. to rank, categorise, discriminate or compare between individuals or data sets, or to study changes) anthropometry such as skinfold thicknesses or limb muscle plus bone areas provides information on regional body composition that will do this with less error than the variables derived by transformation. Reference data for anthropometry are much more voluminous

Table 11.8. *Considerations in choosing a method for assessing the body composition of populations*

Study requirements
Precision and accuracy required
Identity of the measurer
Site of measurements (field or laboratory)
Number to be measured
Whether single or serial measurements are to be made
Whether reference data exist
Apparatus
Cost
Availability
Technical difficulty
Maintenance and calibration
Acceptability to subjects
Ease and comfort
Modesty
Time

These considerations are additional to those discussed earlier in the chapter under 'Choosing a method for estimating body composition'.

Table 11.9. *The feasibility (ease of use) and desirability (information gained) of various body composition methods*

	Feasibility	Desirability
Body weight	5	5
Circumferences	4	4
Skinfolds	3–4	5
Impedance	3	2
Ultrasound	3	2
Infrared	3	1
Body water	2	5
Density	2	5
Photon absorption	2	5
Potassium	1	3
Neutron activation	1	5

Feasibility and desirability scales run from 1 (poor) to 5 (best).
Reproduced with permission from N.G. Norgan (1992) Maternal body composition: methods for measuring short term changes. *Journal of the Biosocial Society*, **24**, 367–77.

and well established than for body composition. Fat patterning may complicate some of these indices but it will do the same for estimations. It might be felt necessary to know the whole-body fat content to calculate energy reserves for an energy balance study, but population methods are hardly appropriate for this purpose and in any case most body composition methods are compromised during changing energy balance.

When it is imperative that estimates of fatness or energy stores or of leanness or muscle mass are obtained, the results should be used cautiously. If transformation is felt to be necessary, the following validated approaches are recommended:

Children: Slaughter *et al.* (1988), Deurenberg *et al.* (1989), Deurenberg *et al.* (1990*b*).
Adults: men: Durnin & Womersley (1974), Jackson & Pollock, (1978); women: Durnin & Womersley (1974), Jackson *et al.* (1980).
Elderly: Chumlea & Baumgartner (1989).

In conclusion, anthropometric variables themselves should be used wherever possible. If transformation is felt to be unavoidable, validated equations should be used. For non-Europeans or populations of non-European origin, Durnin & Womersley's equation has been tried most often and appears the best of a poor bunch. It does not always give accurate population estimates of fat and leanness compared with laboratory methods and provides acceptable individual estimates even less often. BMI and body composition relationships differ between populations. Shape differences may contribute to this and may have an effect in individuals too. BIA and NIRI in particular require further validation before they can be applied to populations other than Europeans. Such simple techniques as visual assessment may meet the needs of some investigators.

References

Alexander, H.G., & Dugdale, A.E. (1992). Fascial planes within subcutaneous fat in humans. *European Journal of Clinical Nutrition*, **46**, 903–6.

Apud, E., & Jones, P.R.M.J. (1980). Validez de la medicion del grosor de los pliegues de grasa subcutanea en estudios de composicion corporal, con referenceia a las ecuaciones de Durnin y Womersley. *Revista Medica de Chile (Santiago)*, **108**, 807–13.

Baumgartner, R.N., Chumlea, W.C., & Roche, A.F. (1990). Bioelectric impedance for body composition. *Exercise and Sports Science Review*, **18**, 193–224.

Baumgartner, R.N., Heymsfield, S.B., Lichtman, S., Wang, J., & Pierson, R.N. (1991). Body composition in elderly people: effect of criterion estimates on predictive equations. *American Journal of Clinical Nutrition*, **53**, 1345–53.

Bland, J.M., & Altman, D.G. (1986). Statistical methods for assessing the agreement between two methods of clinical measurement. *Lancet*, **i**, 307–10.

Boileau, R.A., Lohman, T.G., & Slaughter, M.H. (1985). Exercise and body composition of children and youth. *Scandinavian Journal of Sports Science*, **7**, 17–27.

Brodie, D.A., & Eston, R.G. (1992). Body fat estimations by electrical impedance and infra-red interactance. *International Journal of Sports Medicine*, **13**, 319–25.

Brooke-Wavell, K. (1992). Human body composition: measurement and relationship with exercise, dietary intake and cardio-vascular risk factors. Unpublished PhD Dissertation, Loughborough University of Technology, Loughborough, UK.

Brozek, J., & Keys, A. (1951). Evaluation of leanness-fatness: norms and interrelationships. *British Journal of Nutrition*, **5**, 194–206.

Chumlea, W.C., & Baumgartner, R.N. (1989). Status of anthropometry and body composition data in elderly subjects. *American Journal of Clinical Nutrition*, **50**, 1158–66.

Davies, P.S.W., & Preece, M.A. (1988). Body composition in children: methods of assessment. In *The Physiology of Human Growth* ed. J.M. Tanner & M.A. Preece, pp. 95–107. Cambridge: Cambridge University Press.

Deurenberg, P., van der Kooy, K., Hulshof, T., & Evers, P. (1989). Body mass index as a measure of body fatness in the elderly. *European Journal of Clinical Nutrition*, **43**, 231–6.

Deurenberg, P., Kusters, C.S.L., & Smit, H.E. (1990*a*). Assessment of body composition by bioelectrical impedance in children and young adults is strongly age-dependent. *European Journal of Clinical Nutrition*, **44**, 261–8.

Deurenberg, P., Pieters, J.J.L., & Hautvast, J.G.A.J. (1990*b*). The assessment of the body fat percentage by skinfold thickness measurements in childhood and young adolescence. *British Journal of Nutrition*, **63**, 293–303.

Deurenberg, P., van der Kooij, K., Evers, P., & Hulshof, T. (1990*c*). Assessment of body composition by bioelectrical impedance in a population aged > 60 y. *American Journal of Clinical Nutrition*, **51**, 3–6.

Deurenberg, P., Westrate, J.A., & Seidell, J.C. (1991). Body mass index as a measure of body fatness: age- and sex-specific prediction formulas. *British Journal of Nutrition*, **65**, 105–114.

Diaz, E.O., Vitlar, J., Immink, M., & Czales, T. (1989). Bioimpedance or anthropometry? *European Journal of Clinical Nutrition*, **43**, 129–37.

Diaz, E., Gonzalez-Cossio, T., Rivers, J., Immink, M.D.C., & Dario, R. D. (1991). Body composition estimates using different measurement techniques in a sample of Highland subsistence farmers in Guatemala. *American Journal of Human Biology*, **3**, 525–30.

Durnin, J.V.G.A., & Womersley, J. (1974). Body fat assessed from total body density and its estimation from skinfold thickness: measurements on 481 men and women aged from 16 to 72 years. *British Journal of Nutrition*, **32**, 77–97.

Eckerson, J.M., Housh, T.J., & Johnson, G.O. (1992). The validity of visual estimations of percent body fat in lean males. *Medicine and Science in Sports and Exercise*, **24**, 615–18.

Elia, M., Parkinson, S.A., & Diaz, E.O. (1990). Evaluation of near-infra red reactance as a method for predicting body composition. *European Journal of Clinical Nutrition*, **44**, 113–21.

Ferro-Luzzi, A., Sette, S., Franklin, M., & James, W.P.T. (1992). A simplified approach of assessing adult chronic energy defiency. *European Journal of Clinical Nutrition*, **46**, 173–6.

Friedl, K.E., & Vogel, J.A. (1991). Looking for a few good generalized body fat equations. *American Journal of Clinical Nutrition*, **53**, 795–6.

Fuller, N.J., Jebb, S.A., Laskey, M.A., Coward, W.A., & Elia, M. (1992). Four-component model for the assessment of body composition in humans: comparison with alternative methods, and evaluation of density and hydration of fat-free mass. *Clinical Science*, **82**, 687–93.

Heymsfield, S.B., Wang, J.C., Lichtman, S., Kamen, Y., Kehayias, J., & Pierson, R.N. (1989). Body composition in elderly subjects: a critical appraisal of clinical methodolgy. *American Journal of Clinical Nutrition*, **50**, 1167–75.

Heyward, V.H., Jenkins, K.A., Cook, K.L. (1992). Validity of single-site and multi-site models for estimating body composition of women using near-infrared interactance. *American Journal of Human Biology*, **4**, 579–93.

Hogdon, J., Fitzgerald, P., & Vogel, J. (1989). Visual estimation of fatness. *Medicine and Science in Sports and Exercise*, **21**, S102 (abstract).

Hortobagyi, T., Israel, R.G., Houmard, J.A., O'Brien, K.F., Johns, R.A., & Wells, J.M. (1992). Comparison of four methods to assess body composition in Black and White athletes. *International Journal of Sport Nutrition*, **2**, 60–74.

Houtkooper, L.B., Lohman, T.G., Going, S.B., & Hall, M.C. (1989). Validity of bioelectric impedance for body composition assessment in children. *Journal of Applied Physiology*, **66**, 814–21.

Immink, M.D.C., Flores, R.A., & Diaz, E.O. (1991). Body mass index, body composition and the chronic energy deficiency classification of rural adult populations in Guatemala. *European Journal of Clinical Nutrition*, **46**, 419–27.

Israel, R.G., Houmard, J.A., O'Brien, K.F., McCammon, M.R., Zamora, B.S., & Eaton, A.W. (1989). Validity of a near-infrared spectrophotometry device for estimating human body composition. *Research Quarterly in Exercise and Sports*, **60**, 379–83.

Jackson, A.S., & Pollock, M.L. (1978). Generalised equations for prediciting body density of men. *British Journal of Nutrition*, **40**, 497–504.

Jackson, A.S., Pollock, M.L., & Ward, A. (1980). Generalised equations for predicting body density of women. *Medicine and Science in Sports and Exercise*, **12**, 175–82.

Jackson, A.S., Pollock, M.L., Graves, J.E., & Mahar, M.T. (1988). Reliability and validity of bioelectrical impedance in determining body composition. *Journal of Applied Physiology*, **64**, 529–34.

James, W.P.T., Ferro-Luzzi, A., & Waterlow, J.C. (1988). Definition of chronic energy deficiency in adults. Report of a Working Part of the International Dietary Energy Group. *European Journal of Clinical Nutrition*, **42**, 969–81.

Johnston, F.E. (1982). Relationships between body composition and anthropometry. *Human Biology*, **54**, 221–245.

Jones, P.R.M.J., Bharadwaj, H., Bhatia, M.R., & Malhotra, M.S. (1976). Differences between ethnic groups in the relationship of skinfold thickness to body density. In *Selected Topics in Environmental Biology*, ed. B. Bhatia, G.S. Chhina, & B. Singh, pp. 373–6. New Delhi: Interprint Publications.

Jones, P.R.M.J., Satwanti, K., Bharadwaj, H., Bhatia, M.R., Zachariah, Z., & Kishnani, S. (1982). Relationship of body density and lean body mass to body measurements: application to Indian soldiers of relationships developed for people of European descent. *Annals of Human Biology*, **9**, 355–62.

Katch, F.I. & Katch, V.L. (1980). Measurement and prediction errors in body composition assessment and the search for the perfect prediction equation. *Research Quarterly for Exercise and Sport*, **51**, 249–60.

Katch, F.I., & McArdle, W. (1973). Prediction of body density from simple anthropometric measurements in college age women and men. *Human Biology*, **45**, 445–54.

Kushner, R. F., & Schoeller, D. A. (1986). Estimation of total body water by bioelectric impedance analysis. *American Journal of Clinical Nutrition*, **44**, 417–24.

Lawrence, M., Thongprasert, K., & Durnin, J.V.G.A. (1988). Between-group differences in basal metabolic rates: an analysis of data collected in Scotland, The Gambia and Thailand. *European Journal of Clinical Nutrition*, **42**, 877–891.

Light, R.J., & Pillemer, D.B. (1984). *Summing Up: the Science of Reviewing Research.* Cambridge, Mass: Harvard University Press.

Lohman, T.G. (1981). Skinfolds and body density and their relation to body fatness. *Human Biology*, **53**, 181–225.

Lohman, T.G. (1989). Assessment of body composition in children. *Pediatric Exercise Science*, **1**, 19–30.

Lohman, T.G. (1992). *Advances in Body Composition Assessment*. Champaign, Ill: Human Kinetics.

Lohman, T.G., Pollock, M.L., Slaughter, M.H., Brandon, J., & Boileau, R.A. (1984). Methodological factors and the prediction of body fat in female athletes. *Medicine and Science in Sports and Exercise*, **16**, 182–7.

Lukaski, H.C., Bolonchuk, W.W., Hall, C.B., & Siders, W.A. (1986). Validation of tetrapolar bioelectrical impedance method to assess human body composition. *Journal of Applied Physiology*, **60**, 1327–32.

Minghelli, G., Schutz, Y., Charbonnier, A., Whitehead, R., & Jèquier, E. (1990). Twenty-four-hour energy expenditure and basal metabolic rate measured in a whole-body indirect calorimeter in Gambian men. *American Journal of Clinical Nutrition*, **51**, 563–70.

Nielsen, D.H., Cassady, S.L., Janz, K.F., Cook, J.S., Hansen, J.R., & Wu, J.-T. (1993). Criterion methods of body composition analysis for children and adolescents. *American Journal of Human Biology*, **5**, 211–23.

Norgan, N.G. (1990). Body mass index and body energy stores in developing countries. *European Journal of Clinical Nutrition*, **44** (Suppl 1), 79–84.

Norgan, N.G. (1991). Anthropometric assessment of body fat and fatness. In Anthropometric assessment of nutritional status, ed. J.H. Himes, pp. 197–212. New York: A.R. Liss.

Norgan, N.G. (1994*a*). On the interpretation of low body mass indices:Australian Aborigines. *American Journal of Physical Anthropology*, **94**, 229–37.

Norgan, N.G. (1994*b*). Relative sitting height and the interpretation of the body mass index. *Annals of Human Biology*, **21**, 79–82.

Norgan, N.G. (1994*c*). Population differences in body composition in relation to the Body Mass Index. *European Journal of Clinical Nutrition*, **48** (Suppl. 2).

Norgan, N.G., & Ferro-Luzzi, A. (1982). Weight–height indices as estimators of fatness in men. *Human Nutrition: Clinical Nutrition*, **36**C, 363–72.

Norgan, N.G., & Ferro-Luzzi, A. (1985). The estimation of body density in men: are general equations general? *Annals of Human Biology*, **12**, 1–15.

Norgan, N.G., Ferro-Luzzi, A., & Durnin, J.V.G.A. (1982). The body composition of New Guinean adults in contrasting environments. *Annals of Human Biology*, **9**, 343–53.
Parizkova, J. (1977). *Body Fat and Physical Fitness*. The Hague: Martinus Nijhoff.
Pascale, L.R., Grossman, M.I., Sloane, H.S., & Frankel, T. (1956). Correlations between thickness of skinfolds and body density in 88 soldiers. *Human Biology*, **28**, 165–176.
Pierson, R.N., Wang, J., Heymsfield, S.B. (1991). Measuring body fat: calibrating the rulers. Intermethod comparisons in 389 normal Caucasian subjects. *American Journal of Physiology*, **261** (Endocrinology and Metabolism), E103–8.
Pollock, M.L., Hickman, T., Kendrick, Z., Jackson, A., Linnerud, A.C., & Dawson, G. (1976). Prediciton of body density in young and middle-aged men. *Journal of Applied Physiology*, **40**, 300–4.
Ramirez, M.E. (1992). Measurement of subcutaneous adipose tissue using ultrasound images. *American Journal of Physical Anthropology*, **89**, 347–57.
Ross Conferences on Pediatric Research. (1989). Body composition measurements in infants and children. Report of the 98th Ross Conference on Pediatric Research. Columbus, Ohio: Ross Laboratories.
Satwanti, K., Bharadwaj, H., & Singh, I.P. (1977). Relationship of body density to body measurement in young Punjabi women: applicability of body composition prediction equations developed for women of European descent. *Human Biology*, **49**, 203–13.
Schultink, W.J., Lawrence, M., van Raaij, J.M.A., Scott, W.M., & Hautvast, J.G.A.J. (1992). Body composition of rural Beninese women in different seasons assessed by skinfold thicknesses and bioelectrical impedance measurements and by deuterium oxide dilution technique. *American Journal of Clinical Nutrition*, **55**, 321–25.
Schutte, J.E., Townsend, E.J., Hugg, J., Shoup, R.F., Malina, R.M., & Blomqvist, C.G. (1984). Density of lean body mass is greater in Blacks than in Whites. *Journal of Applied Physiology*, **56**, 1647–9.
Segal, K.R., Van Loan, M., Fitzgerald, P.I., Hodgdon, J.A., & Van Itallie, T.B. (1988). Lean body mass estimation by bioelectrical impedance analysis: a four-site cross-validation study. *American Journal of Clinical Nutrition*, **47**, 7–14.
Sen, R.N., & Banerjee, S. (1958). Studies on the determination of body fat in Indians. *Indian Journal of Medical Research*, **46**, 556–60.
Shephard, R.J., Hatcher, J., & Rode, A. (1973). On the body composition of the Eskimo. *European Journal of Applied Physiology*, **32**, 3–15.
Shetty, P.S., & James, W.P.T. (1994). *Body Mass Index: a Measure of Chronic Energy Deficiency in Adults*. Rome: Food & Agricultural Organization.
Sinning, W.E., Dolny, D.G., Little, K.D. *et al.* (1985). Validity of 'generalised' equations for body composition analysis in male athletes. *Medicine and Science in Sports and Exercise*, **17**, 124–30.
Slaughter, M.H., Lohman, T.G., Boileau, R.A. (1988). Skinfold equations for estimation of body fatness in children and youth. *Human Biology*, **60**, 709–23.
Sloan, A.W. (1967). Estimating body fat in young men. *Journal of Applied Physiology*, **23**, 311–15.
Sparling, P.B., Millard-Stafford, M., Rosskopf, L.B., Dicarlo, L.J., & Hinson, B.T.

(1993). Body composition by bioelectric impedance and densitometry in black women. *American Journal of Human Biology*, **5**, 111–17.

Spurr, G.B., Barac-Nieto, M., Lotero, H., & Dahners, H.W. (1981). Comparisons of body fat estimated from total body water and skinfold thicknesses of undernourished men. *American Journal of Clinical Nutrition*, **34**, 1944–53.

Sterner, T.G., & Burke, E.J. (1986). Body fat assessment: a comparison of visual estimation and skinfold techniques. *Physician and Sports Medicine*, **14**, 101–7.

Ulijaszek, S.J., Lourie, J.A., Taufe, T., & Pumuye, A. (1989). The Ok Tedi Health and Nutrition Project, Papua New Guinea: adult physique of three populations in the North Fly region. *Annals of Human Biology*, **16**, 61–74.

Van Loan, M.D., & Mayclin, P.L. (1987). Bioelectric impedance analysis: is it a reliable estimate of lean body mass and total body water. *Human Biology*, **59**, 299–309.

Vickery, S.R., Cureton, K.J., & Collins, M.A. (1988). Prediction of body density from skinfolds in black and white young men. *Human Biology*, **60**, 135–49.

Wang, Z., Pierson, R.N., & Heymsfield, S.B. (1992). The five-level model: a new approach to organizing body composition research. *American Journal of Clinical Nutrition*, **56**, 19–28.

Westrate, J.A., & Deurenberg, P. (1989). Body composition in children: proposal for a method for calculating body fat percentage from total body density or skinfold thickness measurements. *American Journal of Clinical Nutrition*, **50**, 1104–15.

Willett, W. (1990). *Nutritional Epidemiology*. Oxford: Oxford University Press.

Wilmore, J.H., & Behnke, A.R. (1969). An anthropometric estimation of body density and lean body weight in young women. *American Journal of Clinical Nutrition*, **23**, 267–74.

Womersley, J., & Durnin, J.V.G.A. (1977). A comparison of the skinfold method with extent of overweight and various weight–height relationships in the assessment of obesity. *British Journal of Nutrition*, **38**, 271–84.

Zillikens, M.C., & Conway, J.M. (1990). Anthropometry in blacks: applicability of generalised skinfold equations and differences in fat patterning between blacks and whites. *American Journal of Clinical Nutrition*, **52**, 45–51.

12 *Changes in approach to the measurement of body composition*

JANA PARIZKOVA

Several decades of research in the field of body composition have highlighted a number of new aspects to the subject. The differentiation of body compartments has also become more specific through the development of new methods.

Historical overview

Behnke (1942) examined healthy young men using a two-compartment model which divided the human body into lean body mass and fat mass. He was attempting to characterise young males for élite service in the Marines who had a high body mass index (BMI) but who were at the same time lean and fit American football players. Under such conditions the two-compartment model seemed to be satisfactory, as it assumed that the composition and density of lean body mass and fat were normal and stable. Later measurements have shown that this is not always true – for example in our laboratory we found in some athletes, such as ectomorphic long distance runners or champion hockey players, that total body density was higher than 1.1, which implies differences in the density of lean body mass in these athletes (Parizkova, 1977).

In 1953, Keys & Brozek showed the need to measure the volume of extra-cellular fluid (ECF) in addition to the usual body composition compartments, especially in malnourished subjects, in whom body hydration is disturbed. Subsequently Moore *et al.* (1963) emphasised that body composition research involves the study of biochemical phases, i.e. the fluids and solids that constitute the human body. They used an isotope dilution approach to differentiate the human body into multiple compartments (Table 12.1). Changes in body composition due to various clinical conditions including chronic wasting disease, acute injury and infection, haemorrhage, heart, renal and hepatic disease and obesity could thus be measured. However, this approach was not only costly and demanding for the researchers and their subjects, but it also posed ethical problems due to the invasive nature of the methods used.

Table 12.1. *The multiple-compartment model of the human body of Moore* et al. *(1963)*

The study of body composition is the study of the gross chemical anatomy of the living body. Its approach by the methods of isotope dilution permits the quantification of a wide variety of bodily components. Those listed here were measured by the simultaneous dilution of five isotopes (sodium, bromide, potassium, chromium and deuterium) and the blue dye T-1824. These dilutions yield direct data as follows:

Plasma volume (PV)
Red cell volume (RV)
Blood volume (BV or RV + PV)
Total-body water (TBW)
Extracellular water (ECW)
Total exchangeable sodium (Na)
Total exchangeable potassium (K)

The body weight (B.Wt.) was measured on a Toledo Bed Scale without clothing.

From these direct measurements a number of derivations are derived as follows:

Total-body fat (TBF)
Total-body solids (TBS)
Fat-free solids (FFS)
Fat-free body weight (FFB)
Large vessel haematocrit (LVH)
Whole-body haematocrit (WBH or RV/PR + RV)
Intracellular water (ICW)
Total exchangeable cation or total exchangeable 'base' (Na + K)
Total extracellular sodium (ECN)
Total extracellular potassium (ECK)
Total intracellular potassium (ICK)
Residual sodium (sodium not accounted for in the extracellular water) (ResNa)
Body cell mass (BCM)
Extracellular tissues (ECT)
Interstitial fluid (IF)
Bone (dry bone matrix plus minerals) (B)

With these direct measurements and derivations there may be calculated a wide variety of ratios and relations. Those most commonly referred to are:

Haematocrit ratio (WBH/LVH)
Ratio of total exchangeable sodium to total exchangeable potassium (Na/K)
Ratio of total exchangeable cation to body weight (Na + K/B.Wt)
Ratio of total exchangeable cation to total-body water (Na + K/TBW)
Ratio of intracellular water to total-body water (ICW/TBW)
Ratio of red cell volume to total exchangeable potassium (RV/K)
Ratio of total exchangeable potassium to dry body weight (K /DBW)
Ratio of total exchangeable potassium to fat-free solids (K /FFS)
Ratio of red cell volume to plasma volume (RV/PV)
Ratio of plasma volume to total exchangeable sodium (PV/Na)

After Moore *et al.* (1963) *The Body Cell Mass and its Supporting Environment*. Philadelphia: W.B. Saunders.

Recent developments

Substantial changes in the approach to body composition evaluation have occurred in recent years. The five-level model of body composition has been suggested: atomic, molecular, cellular, tissue system and whole body (Heymsfield *et al.*, 1993). Components at each level have been separately defined, although there are associations between the levels.

Until recently, the two-compartment model of body composition has assumed that lean body mass and fat are constant in composition. However, for studies covering the entire life span it is necessary to use techniques that are not affected by ageing. Present techniques, both traditional and modern, can, according to Kehayias (1993), measure body fat on the basis of the following five principles:

1. Measurement of a physical property of the body such as density, impedance, photon attenuation and electrical conductivity. The corresponding techniques are densitometry, bioelectrical impedance, dual-photon absorptiometry and total body conductivity (TOBEC). Anthropometric measurements including BMI are included in the evaluation.
2. Measurement of an index of lean body mass via measurements such as total body water (TBW), total body potassium (K) and nitrogen (N). The measured quantity is then adjusted to represent lean body mass and, by subtraction, body fat.
3. Measurements of several components of lean body mass which require at least one neutron activation technique. Some of the best models have been developed at the Brookhaven National Laboratory (Cohn *et al.*, 1984). They monitor the main components of
   LBM I = TBW + protein + bone mass, or
   LBM II = BCM + ECF + ECS
   (TBW by tritiated or heavy water dilution, body cell mass (BCM) by counting natural radioactivity of body potassium, ECF by total body chlorine, extracellular solids (ECS) by body calcium (Ca) and chlorine (Cl) by delayed neutron activation analysis: Cohn *et al.*, 1984). Protein is measured via total body nitrogen, using prompt gamma neutron activation (Vartsky *et al.*, 1979, 1982).
4. Imaging techniques, which include application of computed tomography (CT) and magnetic resonance imaging (MRI). These methods focus mainly on localising and quantifying body fat and recognising patterns of adipose tissue distribution.
5. The elemental partition analysis approach which has resulted from the introduction of neutron inelastic scattering (Kehayias, 1993) and neutron capture techniques. This has led to a new 'assumption-

free' approach called elemental partition analysis in which the major elements of the body (carbon, hydrogen (H) and nitrogen) are measured and partitioned into the contributing body compartments.

Then there exist further models such as the C–N–Ca model, the C–K model and the H–TBW model among others. Techniques under development are a neutron inelastic scattering method for the measurement of regional body fat stressing the importance of the carbon-to-oxygen ratio. Gamma ray resonance absorption analysis is another technique which has the advantage of a very low radiation exposure to the subjects during the measurement of nitrogen, oxygen and carbon (Kehayias, 1993).

The comparison of traditional and advanced technologies

Many of the newer methods have been compared with classical methods such as densitometry and the measurement of total body water, as well as with classical anthropometry. Ultrasound methods, TOBEC, bioimpedance analysis and near-infrared interactance (NIRI) have also been tested in a similar way (Cochran *et al.*, 1988). Further studies are devoted to comparisons of these methods (e.g. Hoergenroeder *et al.*, 1991; Forbes, 1987 Shephard *et al.*, 1991; Lohman *et al.*, 1992; Hortobagyi *et al.*, 1992, etc.).

Special attention has also been paid to body composition in childhood (Klish *et al.*, 1987; Fiorotto *et al.*, 1987; Cochran *et al.*, 1988; Fiorotto & Klish, 1991).

It can be seen that two types of study are now widespread: clinical laboratory studies using advanced technologies in a small samples, and field studies using simple classic and/or more recent methods in larger population samples. The latter studies are concerned mostly with the relation between growth and development, and nutritional and health status, physical fitness and work performance, and are often conducted under unfavourable environmental and low budget conditions. Both lines of research have made theoretical and practical contributions to the subject.

Body composition and functional characteristics

One of the most important aspects of the numerous body composition studies has been the identification and definition of the relationships between body compartments and other functional, nutritional, work performance and health-related parameters. Some relationships have been shown to be due to constitutional and genetic factors; others have been established on the basis of adaptive processes during various periods of life (especially during growth), or by a combination of both factors. Such studies have had practical aims, such as to define optimal body composition from the point of view of 'positive health', work performance, definition of recommended dietary allowances, etc. Many of these problems have not yet been satisfactorily answered.

Another change in the approach to the evaluation of body composition has been the focusing of attention not only on the absolute and relative amounts of body compartments, but also on their distribution in different regions of the body. For example CT and/or dual energy X-ray absorptiometry (DXA) are used for all tissues, while subcutaneous fat thickness is measured by skinfold calipers, ultrasound techniques and NIRI.

Genetic impact on body fat and fat patterning during early growth

Data about changes in body composition during growth, obtained using densitometry and skinfold thickness measurements, were first published by Parizkova (1961). Skinfolds were measured during the first 48 h in normal, full-term infants, then in premature infants, and also in infants from diabetic mothers before they became metabolically compensated. It was interesting that not only the total amount of fat but also the fat pattern expressed by various ratios of trunk and extremity skinfolds varied widely in these newborns (Parizkova, 1963). In term newborns the centrality index (subscapular/triceps skinfold ratio) was equal to 1.3, while in premature infants with little subcutaneous fat the index was 1.0, and in children from diabetic mothers with a large deposit of fat the index equalled 1.07, i.e. in both the latter cases the fat pattern was disturbed.

Fat distribution in newborns shows the impact of genetic factors. Alberti Fidanza & Fidanza (1986) assessed anthropometric, biochemical, haematological and nutritional characteristics in mothers during the first, second and third trimesters of pregnancy, and compared them with the same values in their newborns. In a later study we showed significant relationships between maternal biceps, subscapular and calf skinfolds during pregnancy on the one hand, and infant subscapular and calf skinfolds on the other (J. Alberti & J. Parizkova, unpublished data). The correlations between subscapular/triceps ratio and other indices in the mother and the same indices in the newborn were always significant. This relationship was closer in boys (Table 12.2).

We also found significant correlations between the values for some skinfolds of the mother and those measured in her pre-school child. Fat pattern was also correlated at this age (Table 12.3) (Parizkova, unpublished).

Table 12.2. *Correlation coefficients between indices of fat distribution in the mother measured at different trimesters of pregnancy and the same indices measured in the newborn*

	Newborns	
Mothers	Index 1	Index 2
Newborn boys and girls ($n = 44$)		
Second trimester		
Index 1	0.32*	–
Index 2	0.33*	–
Third trimester		
Index 1	0.36*	–
Index 2	0.41**	0.33*
Newborn boys only ($n = 18$)		
First trimester		
Index 1	–	–
Index 2	0.48	–
Second trimester		
Index 1	0.71**	0.69**
Index 2	0.60**	0.64**
Third trimester		
Index 1	0.72**	0.72**
Index 2	0.63**	0.65**

Index 1 (centrality index), subscapular/triceps skinfold; index 2, subscapular/(triceps + biceps) skinfold.
*$p < 0.05$; **$p < 0.01$.

Stability of body composition and development of fat patterning

In boys followed longitudinally from 11 to 18 years of age, body composition shows a stable trend of development indicating a genetic component. Relative amounts of lean body mass and fat correlated significantly after age 5 years, but not after 8 years. Also the correlation coefficients between individual skinfold thicknesses were lower after 8 years, but remained significant (Parizkova, 1977).

The centrality index showed significant changes during this growth period: it did not show the same trend as the percentage of body fat measured by densitometry, i.e. trunk fat increased more than extremity fat in spite of the reduction in the percentage of total-body fat. Another index including more trunk and extremity skinfolds increased slightly but not significantly: it was related significantly only to the absolute amount of body fat (Parizkova, 1993).

The impact of genetic factors and characteristic ontogenetic trends is obvious throughout life, as shown not only by this longitudinal study but also by comparisons of the centrality index with other characteristics such

Table 12.3. *Correlations between anthropometric measures of mothers and their children (2–5 years of age)*

Mothers	Male + female ($n = 72$)	Male ($n = 43$)	Female ($n = 29$)
Height	0.42**	0.57**	0.16
Sitting height	0.61**	0.78**	0.34
Biacromial diameter	0.41**	0.55**	0.07
Thigh circumference	0.22	0.17	0.39*
Skinfolds			
Chin	0.27*	0.46**	−0.12
Thigh	0.22	0.34**	0.05
Calf	0.41**	0.30	0.42
Fat patterning			
Centrality Index	0.66**		
Body mass index	0.69**		

Centrality index, subscapular/triceps skinfold; body mass index, weight/height^2.
*$p < 0.05$; **$p < 0.01$.

as BMI and the percentage of body fat in individual age groups. BMI correlated significantly with total-body fat as measured by densitometry (Parizkova, 1989). Correlation of BMI and the percentage of body fat helped to characterise various groups of youths and adults, enabling a more detailed definition of body physique to be made.

Body composition and fat distribution changes in obese children after treatment to reduce weight

Longitudinal studies of obese children on reduction therapy showed a decrease in BMI and percentage body fat, along with improvements in functional capacity. Longitudinal observations of the sum of ten skinfolds showed that, so long as the regime was adhered to, fat continued to reduce after treatment (Parizkova, 1993; Parizkova & Hainer, 1990). At the same time there were significant changes in fat pattern related to different periods of treatment, and they gradually approached standard values for normal children of the same age as regards body composition, fat patterning and aerobic capacity (Parizkova, 1993).

Body composition, enzymatic activity and muscle fibre pattern

A relationship between body composition and the level of functional capacity and physical fitness has been shown in many studies. Certain types of physique may predispose individuals towards certain athletic activities:

on the other hand, certain types of exercise may significantly change body composition with the development of lean body mass at the expense of fat, along with increased aerobic power in dynamic sport disciplines.

There is interest therefore in the relationship between body components and other characteristics of skeletal muscle such as enzymatic activities and/or muscle fibre pattern. Comparison of the percentage of body fat, maximal oxygen uptake (characterising the aerobic power), and the activities of hydroxyacylCo-A dehydrogenase, malate dehydrogenase (MDH) and citrate synthetase in biopsies (vastus lateralis muscle) in skiers on very high or low levels of performance showed significant differences between the groups. Lean skiers with the higher level of maximal oxygen uptake also had higher levels of activity for the above mentioned enzymes. There was a significant negative relationship between the percentage of body fat and the activity of hydroxyacylCo-A dehydrogenase: the lower the percentage fat, the higher the level of hydroxyacylCo-A dehydrogenase. Similar correlations were found for MDH and lactate dehydrogenase (Parizkova *et al.*, 1987). Such relationships could be shown only in athletes adapted to dynamic aerobic exercise, not in normal subjects. We may assume that the adaptation to exercise simultaneously changed both characteristics.

The same pattern applied to the relationships between body compartments and muscle fibre pattern, i.e. the ratio of slow-oxidative (SO; fibre type I), fast-oxidative glycolytic (FOG; type IIa) and fast-glycolytic (FG; type IIb) muscle fibres taken from the vastus lateralis muscle by biopsy. The relationship was significant and more apparent in the group of athletes than in the total group including control males (Parizkova, 1993). We assume that this relationship appeared due to the selection of athletes predisposed to dynamic exercise by a higher proportion of SO muscle fibres, who kept a low level of body fat by dynamic, aerobic training (Melichna *et al.*, 1994).

Body composition and the adaptation to cold

A group of ten students from the Faculty of Physical Education in Prague was adapted to cold by immersion for 1 h in water at a temperature of 14 °C three times per week. The first experiment lasted 4 weeks, and the second 6 weeks.

The adaptation was both isolatory and hypothermic, and was manifested by a decrease in the thermoregulatory set-point. Peripheral temperature was lowered as a result of accentuated vasoconstriction; central temperature was lowered too, resulting from a later start of muscle tremor. During a 1 h session in the cold water, energy output decreased by 20% (Jánský *et al.*, 1992).

The sum of ten skinfolds increased slightly but significantly during the

Table 12.4. *Morphological and nutritional changes in 10 subjects after 4 weeks of adaptation to cold*

	Before		After	
	Mean	SD	Mean	SD
Height (cm)	181.1	3.4	–	–
Weight (kg)	75.3	2.4	75.3	2.8
BMI (kg/m^2)	23.0	0.9	23.0	0.9
Sum of 10 skinfolds (mm)	56.9	4.4	64.6	4.9*
Sum of 5 skinfolds (mm)	30.6	1.8	32.8	2.0*
Dietary intake				
Energy (MJ)	14.6	2.8	17.7	2.2*
Protein, total (g)	116.0	20.0	134.3	16.0*
Protein, animal (g)	54.3	9.7	64.7	4.6*
Protein, plant (g)	61.7	11.8	69.6	4.6*
Fat, total (g)	126.0	25.3	134.0	33.7*
Fat, animal (g)	82.0	13.2	84.7	19.3*
Fat, plant (g)	44.0	20.4	49.3	25.8*

*$p < 0.05$.

experiment. There was also an increase in energy, protein, fat and carbohydrate intake (measured by the inventory method) during 1 week at the beginning and end of the first experiment. Measurements of densitometry and bioelectrical impedance in the second experiment did not show significant changes in body composition (Table 12.4). The latter method gave lower values of body fat and did not correlate with densitometry. However, skinfold thickness increased significantly for some parts of the body. Neither fat pattern nor aerobic power changed in either experiment. In the second experiment the nutritional intake data showed only insignificant changes after 6 weeks (Table 12.5) (Parizkova *et al.*, unpublished data).

Body composition, enzyme activity and muscle fibre characteristics in vegetarians

The impact of a vegetarian diet on body composition was studied in another group of subjects. In a preliminary study of a larger sample of vegetarians ($n = 56$) BMI and percentage of body fat were significantly lower than in a control group. In the smaller sample ($n = 14$) the somatic characteristics did not show significant differences (Table 12.6) that could be related to diet composition (Table 12.7) or muscle enzyme activity (Table 12.9). The percentage of FG, FOG and SO fibres in the vastus lateralis muscle did not show significant differences between the groups except for SO fibres in females (Table 12.8). Triosophosphate dehydrogenase

Table 12.5. *Morphological and nutritional changes in 10 subjects after 6 weeks of adaptation to cold*

	Before		After	
	Mean	SD	Mean	SD
Height (cm)	184.3	4.34		
Weight (kg)	78.7	6.4	79.7	6.4
BMI (kg/m^2)	23.2	2.1	23.5	2.1
Fat (%)	13.7	2.1	14.3	3.0
Skinfolds				
Cheek (mm)	6.3	1.4	6.9	1.4*
Chin (mm)	3.5	2.0	3.5	1.7
Thorax I (mm)	1.8	0.5	2.5	0.8*
Triceps (mm)	6.4	2.9	7.1	2.8*
Subscapular (mm)	9.3	3.0	9.8	3.3
Thorax II (mm)	6.0	4.4	6.5	4.3
Abdomen (mm)	11.3	5.3	13.4	6.4**
Suprailiac (mm)	6.9	3.8	8.1	4.6
Thigh (mm)	6.2	1.6	7.8	2.4*
Calf (mm)	5.8	2.3	7.1	3.0**
Dietary intake				
Energy (MJ)	15.3	4.9	17.1	3.5
Protein (g)	129.5	32.6	124.6	16.1
Fat (g)	134.0	61.0	157.6	48.5
Carbohydrate (g)	518.5	170.0	514.9	93.5

*$p < 0.05$; **$p < 0.01$.

and lactate dehydrogenase activity was significantly lower in male vegetarians as compared with controls (Table 12.9). Significant correlations of weight and lean body mass with muscle enzyme activity were found, and those for lean body mass were higher than for weight. Dietary intake correlated with lean body mass but not with weight (Table 12.10) (J. Parizkova *et al.*, unpublished data).

Body composition and functional changes in adult obesity after reduction treatment

Studies in the adult obese after 4 weeks of reduction treatment showed a substantial decrease in body weight, BMI and body fat, but also some reduction in lean body mass (Parizkova & Hainer, 1990). Blood pressure, blood cholesterol and triglycerides also decreased significantly. The waist/hip ratio did not change and neither did the centrality index (subscapular/triceps skinfold ratio). In the chronically obese the response to treatment (a very low energy diet and mild, tolerable exercise) did not change the fat pattern,

Table 12.6. *Morphological characteristics of vegetarian and control subjects*

	Males				Females			
	Vegetarians ($n = 7$)		Controls ($n = 6$)		Vegetarians ($n = 7$)		Controls ($n = 7$)	
	Mean	SD	Mean	SD	Mean	SD	Mean	SD
Age (years)	34.9	6.0	35.6	6.6	33.1	8.4	32.0	9.9
Height (cm)	181.6	4.7	180.5	7.1	167.7	7.9	166.1	4.1
Weight (kg)	74.6	10.6	76.4	8.6	61.6	10.9	61.2	12.7
BMI (kg/m^2)	22.6	2.9	23.3	1.0	21.7	2.4	22.1	4.0
Fat (%)	11.9	5.2	9.9	3.8	17.5	5.3	17.7	6.8
LBM (kg)	65.7	6.1	68.5	5.4	50.4	7.0	48.8	7.8
Centrality index I	1.03	0.21	1.12	0.26	0.64	0.15	0.56	0.31
Centrality index II	1.33	0.32	1.29	0.20	1.04	0.13	0.83	0.29

Table 12.7. *Food intake in vegetarian and control subjects*

	Males				Females			
	Vegetarians		Controls		Vegetarians		Controls	
	Mean	SD	Mean	SD	Mean	SD	Mean	SD
Energy (MJ)	12.7	5.1	15.0	3.7	7.8	1.1*	11.5	2.4
Protein total (g)	99.9	34.5	122.5	30.0	56.8	11.2*	96.7	24.5
Animal	17.5	12.6*	63.9	18.0	12.2	7.9*	49.7	14.1
Plant	82.4	39.0	58.6	25.8	44.6	8.5	47.0	14.4
Fat total (g)	69.3	29.2*	132.3	32.6	46.9	13.8*	101.9	25.5
Animal	21.7	13.1*	81.1	25.4	15.8	8.0*	66.5	25.7
Plant	47.6	21.5	51.2	23.3	31.1	11.4	35.1	9.8
Carbohydrate (g)	513.5	226.7	454.7	160.2	326.1	55.7	347.4	74.9

*$p < 0.05$.

Table 12.8. *Percentage of different muscle fibres (vastus lateralis muscle biopsy) in vegetarian and control subjects*

	Males				Females			
	Vegetarians		Controls		Vegetarians		Controls	
	Mean	SD	Mean	SD	Mean	SD	Mean	SD
FG	16.9	7.9	20.7	10.8	17.6	10.7	26.1	9.6
FOG	30.3	8.5	28.5	13.4	28.7	11.6	38.9	9.9
SO	52.8	12.6	35.8	17.0	53.6	6.2*	35.0	10.1
FG + FOG	47.2	12.6	49.2	23.5	46.3	6.3*	65.0	10.1

FOG, fast-oxidative glycolytic, type IIa muscle fibre; FG, fast-glycolytic, type IIb muscle fibre; SO, slow oxidative, type I muscle fibre.
*$p < 0.05$.

Table 12.9. *Enzyme activity of muscle fibres (vastus lateralis muscle biopsy) in vegetarian and control subjects*

	Males				Females			
	Vegetarians		Controls		Vegetarians		Controls	
	Mean	SD	Mean	SD	Mean	SD	Mean	SD
TPDH	260.0	44.0*	337.3	42.8	254.3	55.6	251.0	63.4
LDH	185.3	60.3*	307.0	83.0	121.2	24.0	124.0	47.0
CS	10.2	5.3	8.6	0.7	8.3	1.0	8.2	2.1
HAD	3.3	2.2	4.2	1.0	4.6	1.2	5.1	0.6
Enzymes (sum.)	459.0	86.0*	657.0	95.5	388.0	65.0	388.0	100.0

TPDH, triosophosphate dehydrogenase; LDH, lactate dehydrogenase; CS, citrate synthetase; HAD, hydroxyacylCoA-dehydrogenase.
*$p < 0.05$.

Table 12.10. *Correlation coefficients of weight and lean body mass with dietary intake and muscle enzyme activity*

	Lean body mass	Body weight
Dietary intake		
Energy	0.44*	–
Protein	0.43*	–
Carbohydrate	0.45*	–
Vitamin B	0.46*	–
Enzyme activity		
TPDH	0.56**	0.50**
LDH	0.59**	0.44*

TPDH, triosophosphate dehydrogenase; LDH, lactate dehydrogenase.
*$p < 0.05$; **$p < 0.01$.

and BMI and fatness remained high after treatment; this required the treatment to be extended. Most of the subjects were followed up in the outpatient department. The centrality index tended to be higher in those who maintained their weight reduction compared with those who regained weight (Hainer *et al.*, 1992).

A similar trend was observed in two groups of obese women with diffuse and/or disproportional 'spider-like' obesity, i.e. gynoid and android types of obesity. The centrality index, and another index of trunk and extremity skinfolds, showed a significantly different distribution of subcutaneous fat with the same amount of total body fat as measured by densitometry (Skamenova & Parizkova, 1963; Parizkova, 1993).

Thus, repeated measurements of total and subcutaneous fat helped to define different types of obesity and their treatment. However, in cases of morbid obesity with additional clinical complications, the two-compartment model was not satisfactory because of abnormal hydration, changed composition of some tissues, etc. For such purposes four- or even six-compartment models would be more suitable for classifying body composition changes before and after weight reduction.

Extreme work load and body composition

It is also necessary to use the more detailed models to measure body composition changes in endurance sports events such as 24 h runs or triathlons. Males with such work loads lost an average of 2–3 kg in weight, but did not change their skinfold thickness. They also showed significant changes in serum triglycerides, total and high density lipoprotein (HDL)-cholesterol, free fatty acids, urea and uric acid, indicating the mobilisation and utilisation of various energy resources necessary for their performance.

There was a remarkable deficit between the dietary energy intake during and immediately after the competitions, and the energy output estimated by indirect calorimetry (Parizkova & Novak, 1991). Obviously water loss contributed significantly to the weight decrement, even though the competitors had free access to drinks. The composition of weight decrements regarding glycogen, fat, protein and water during and after competitions can only be conjectured, as can how and when the situation eventually normalised.

Conclusions

The studies described above, and many others, have stimulated interest in the development of techniques to measure body composition. The application of more sophisticated methods has been important to this process. However, the use of simpler classical methods based on only two compartments of the human body may still be useful for certain research problems, especially when validated by other morphological, functional and biochemical assessments. Thus new problems may be tackled even when advanced technology is not available.

References

Alberti Fidanza, A. & Fidanza, R. (1986). A nutrition study involving a group of pregnant women in Assisi, Italy. *International Journal of Vitamin and Nutrition Research*, **56**, 373–80.

Behnke, R. (1942). Physiologic studies pertaining to deep-sea diving and aviation, especially in relation to the fat content and composition of the body. *Harvey Lecture Series*, **37**, 198–226.

Cochran, W.J., Wong, W.W., Fiorotto, M.L., Sheng, Hwai-Ping, Klein, P.D. & Klish, W.J. (1988). Total body water estimated by measuring total body electrical conductivity. *American Journal of Clinical Nutrition*, **48**, 946–50.

Cohn, S.H., Vaswani, A.N., Yasamura, S., Yuen, K. & Ellis, K.J. (1984). Improved models for determination of body fat in vivo neutron activation. *American Journal of Clinical Nutrition*, **40**, 255–9.

Fiorotto, M.L. & Klish, W.J. (1991). Total body electrical conductivity measurement in the neonate. *Clinics in Perinatology*, **18**, 611–27.

Fiorotto, M.L., Cochran, W.J., Funk, R.C., Sheng, Hwai-Ping. & Klish, W.J. (1987). Total body electrical conductivity measurements: effects of body composition and geometry. *American Journal of Physiology*, **252**, (Regulatory Integrative Comparative Physiology 21). R794–R800.

Forbes, G.W. (1987). *Human Body Composition: Growth, Aging, Nutrition and Activity*. New York: Springer-Verlag.

Hainer, V., Stich, V., Kunesova, M., Parizkova, J., Zak, A., Wernischova, V. & Hrabak, P. (1992). Effect of four-week treatment of obesity by very low calorie diet on anthropometric, metabolic and hormonal indices. *American Journal of Clinical Nutrition*, **56**, 218S–282S.

Heymsfield, S.W., Wang, Z., Baumgartner, R.N., Dilmanian, F.A., Ma, R. & Yasumura, S. (1993). Composition and aging: a study by *in vivo* neutron activation analysis. *Journal of Nutrition*, **123** (Suppl. 2) 432–7.

Hoergenroeder, A.C., Wong, W., Fiorotto, M.L., O'Brien Smith, E. & Klish, W.J. (1991). Total body water and fat-free mass in ballet dancers: comparing isotope dilution and TOBEC. *Medicine and Science in Sports and Exercise*, **23**, 534–41.

Hortobagyi, T., Israel, R.G., Houmard, J.A., McCammon, M.R. & O'Brien, K.F. (1992). Comparison of body composition assessment by hydrodensitometry, skinfolds and multiple site near infrared spectrophotometry. *European Journal of Clinical Nutrition*, **46**, 205–11.

Jánský, L., Hošek, V., Janáková, H., Uličný, B. & Parizkova, J. (1992). Human adaptation to cold: shift of the shivering threshold during repeated water immersion. *Physiological Research*, **41** (Suppl.), 198.

Kehayias, J.J. (1993). Aging and body composition: possibilities for future studies. *Journal of Nutrition*, **123** (Suppl 2), 454–8.

Keys, A. & Brozek, J. (1953). Body fat in adult man. *Physiological Reviews*, **33**, 245–345.

Klish, W.J., Cochran, W.J., Fiorotto, M.L., Wong, W.W. & Klein, P.D. (1987). The bioelectrical measurement of body composition during infancy. *Human Biology*, **59**, 319–27.

Lohman, T.G. (1992). *Advances in Body Composition Assessment*. Champaign, Ill.: Human Kinetics Publishers.

Melichna, J., Parizkova, J., Zauner, C.W. & Havlickova, L. (1994). Relationship of muscle fibre distribution to body composition in physically trained and normally active human males. *Physiological Research*, **43**, 233–41.

Moore, F.D., Olesen, K.H., McMurray, J.D., Parker, H.V., Ball, M.R. & Magnus Boyden, C. (1963). *The body cell mass and its supporting environment: Body Composition in Health and Disease*, Philadelphia: Saunders Company.

Parizkova, J. (1961). Total body fat and skinfold thickness in children. *Metabolism*, **10**, 794–802.

Parizkova, J. (1963). The impact of age, diet and exercise on man's body composition. *Annals of New York Academy of Sciences*, **110**, 661–74.

Parizkova, J. (1977). *Body Fat and Physical Fitness: Body Composition and Lipid Metabolism in Different Regimes of Physical Activity*. The Hague: Martinus Nijhoff.

Parizkova, J. (1989). Age-dependent changes in dietary intake related to work output, physical fitness and body composition. *American Journal of Clinical Nutrition*. **49**, 962–7.

Parizkova, J. (1993). Obesity and its treatment by exercise. In *Nutrition and Fitness in Health and Disease*. World Review of Nutrition and Dietetics vol. 72, ed. A.P. Simopoulos, pp. 78–91. Basel: Karger.

Parizkova, J. & Hainer, V. (1990). Exercise in growing and adult obese individuals. In *Current Therapy in Sports Medicine*, vol. 2, ed. J.S. Torg, R.P. Welsh & R.J. Shephard, pp 22–6.

Parizkova, J. & Novak, J. (1991). Dietary intake and metabolic parameters in adult men during extreme work load. In *Impacts on Nutrition and Health*. World Review of Nutrition and Dietetics, ed. A.P. Simopoulos, pp 72–98. Basel: Karger.

Parizkova, J., Bunc, V., Sprynarova, S., Mackova, E. & Heller, J. (1987). Body composition, aerobic capacity, ventilatory threshold and food intake in different sports. *Annals of Sports Medicine*. **3**, 171–7.

Shephard, R.J. (1991). *Body Composition in Biological Anthropology*. Cambridge: Cambridge University Press.

Skamenova, B. & Parizkova, J. (1963). Assessment of disproportional (spiderline) and diffuse type of obesity by measurement of total and subcutaneous body fat (in Czech). *Casopis lekaru ceskych*. **102**, 142–6.

Vartsky, D., Ellis, K.J. & Cohn, S.H. (1979). *In vivo* measurement of body nitrogen by analysis of prompt-gammas from neutron capture. *Journal of Nuclear Medicine*, **20**, 1158–65.

Vartsky, D., Wiepolsky, L., Ellis, K.J. & Cohn, S.H. (1982). The use of nuclear resonant scattering of gamma rays for in vivo measurement of iron. *Nuclear Instruments Methodology*, **193**, 359–64.

13 *Multi-compartment models for the assessment of body composition in health and disease*

SUSAN A. JEBB AND M. ELIA

Introduction

Body composition analysis has evolved from the single unit of body weight, through the classical division into fat and fat-free mass and then onto three, four or more compartments. This chapter will describe each model in terms of the principle of the method, the compartments which are measured, the techniques required and the theoretical limitations or advantages that the model presents. A summary of each model is given in Table 13.1. Alternative approaches to the division of the body including elemental analysis, intra- and extracellular constituents and body segments will not be addressed further here.

Weight: a single-compartment model?

Although weight alone gives no indication of body composition, in adults it is not uncommon for judgements to be made on the basis of weight, particularly with reference to obesity. There is still a tendency to say that men greater than 100 kg or women greater than 80 kg are overweight with the implied assumption that they have excess body fat. The limitations of this approach are obvious.

Two-compartment models

In the two-compartment model the body is divided into fat and fat-free mass. At its simplest the measurement of weight and height gives some indication of body composition, by reference to tables of ideal body weight, although this again makes the assumption that the majority of excess weight is fat.

The two-compartment model is also the basis of most classical body composition techniques, notably the measurement of body density, total-body water and total-body potassium.

Table 13.1. *Principles and assumptions of different body composition models*

One compartment	Two compartments	Three compartments	Four compartments	Further compartments
Weight	Fat and FFM	Fat, water, protein + mineral	Fat, water, protein, mineral	
'Excess' weight relative to a reference weight, assumed to be fat. Independent of height, muscularity or hydration status of individual	Densitometry: Assumes density of fat and FFM remain constant, typically density of fat = 0.9 kg/l and density of fat-free mass = 1.1 kg/l Total body water (deuterium dilution): Assumes hyration fraction of FFM is known and constant, typically 0.73. Total body potassium: Assumes concentration of potassium in FFM is known and constant, typically 66 mmol/kg (male) and 60 mmol/kg (female)	Assumes: D_2O space/1.04 = TBW Density of fat = 0.9 kg/l Ratio of protein to mineral fixed, such that density of protein + mineral = 1.52 kg/l	Method 1: D_2O space/1.04 = TBW Density of fat = 0.9 kg/l Density of mineral = 3.075 kg/l Density of protein = 1.34 kg/l Method 2: D_2O space/1.04 = TBW TBN × 6.25 = Protein TBCa = 34% bone mineral	As for four-compartment model, plus: Division of TBW into intra- and extracellular compartments. Glycogen calculated as 0.044 g/kg protein Further elemental analysis using IVNAA

FFM, fat-free mass; D_2O, deuterium oxide; TBW, total-body water; TBN, total-body nitrogen; TBCa, total-body calcium; IVNAA, *in vivo* neutron activation analysis.

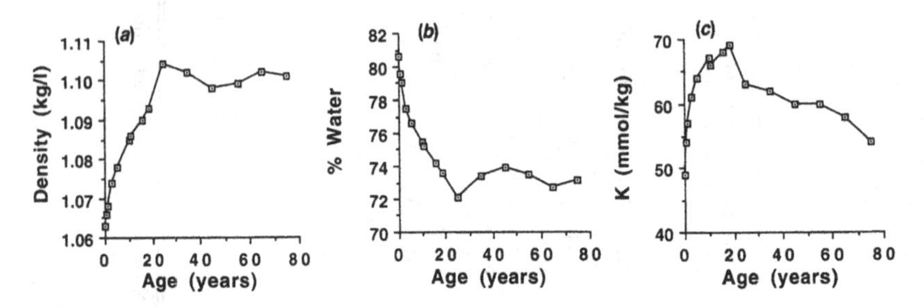

Fig. 13.1. Changes in the composition of fat-free mass with age in Caucasian males: (*a*) density, (*b*) hydration, (*c*) potassium. Based on data from Elia (1992).

Density

Body density is calculated as the mass of the body in air divided by its volume. The measurement of body volume can be achieved by measuring the volume of water displaced when the subject is immersed underwater, or as the difference between the mass in air and the mass underwater. It is then assumed that the densities of the fat and fat-free mass compartments are known and constant. Typically the density of fat is assumed to be 0.9 kg/l and fat-free mass 1.1 kg/l. Hence knowing total-body density the proportion of each of the two compartments, fat and fat-free mass, may be calculated. There is little controversy over the assumption of a constant density of fat but more debate about the density of fat-free mass. If the density of fat-free mass is to remain constant at 1.1 kg/l then its proportional composition must remain constant at all times, including normal growth and development, senescence and disease. In reality this is not the case, because of the variable contributions of the individual constituents of fat-free mass in different situations. Thus the density of fat-free mass varies with age, growth and development, race and disease. The density of fat-free mass is greater in Negroid races who have thick, dense bones; in contrast the density is reduced in patients with osteoporosis or oedema. Fig. 13.1*a* shows the change in the density of fat-free mass with age in Caucasian males, which is due in part to changes in hydration and the bone mineral content of fat-free mass.

Total-body water (TBW)

Nowadays total-body water is most commonly measured using the deuterium oxide isotopic dilution technique. The deuterium space is calculated from the dose of isotope administered and the measured concentration in body water at equilibrium. Since deuterium also exchanges with some non-aqueous hydrogen in protein and carbohydrate it is necessary to make a correction for this sequestration, estimated to be 4% (the reliability of this estimate is one potential source of error). Fat-free mass is then calculated based on the assumption that it has a known hydration fraction (typically assumed to be 73% in adults) and fat is calculated as the difference between fat-free mass and body weight.

However, like the density of fat-free mass, the hydration fraction of lean tissue also varies during normal growth and development, particularly during childhood. From birth to 20 years of age it falls from 81% to 73% (Fig. 13.1*b*). To some extent these changes may be accounted for in the calculation of body composition from total-body water measurements, since an estimate can be made of the most appropriate hydration fraction. However, there are substantial differences between individuals and in the same individual at two or more points in time – hence the impossibility of defining the correct hydration factor for any given isotope dilution measurement. The problems become even greater when assessing the body composition of individuals with abnormal fluid balance (e.g. in pregnancy and in oedematous or dehydrated patients).

Total-body potassium (TBK)

The naturally occurring radioactive isotope of potassium, ^{40}K, can be measured in a whole-body counter and used to derive an estimate of total-body potassium. Although the potassium content of lean tissue varies between males and females, by assuming a constant concentration of potassium in lean tissue, typically 60 mmol/kg for women and 66 mmol/kg for men, total-body potassium measurements can be used to derive an estimate of fat-free mass (Forbes *et al.* 1961). Fat, which is devoid of potassium, is then calculated by subtracting fat-free mass from body weight. The difference between the potassium concentration of fat-free mass between men and women is largely explained by the greater proportion of muscle in men, which has a higher concentration of potassium than non-muscular tissues.

In addition, like density and hydration, potassium concentration varies during growth and development (Fig. 13.1*c*), due mainly to a reduction in the ratio of extracellular water (which is poor in potassium) to intracellular water (which is rich in potassium). It can change even more dramatically in disease states associated with either oedema, where there is an increase in extracellular water, or a depletion of intracellular potassium, such as aggressive and prolonged use of diuretics.

Three-compartment models

In 1961 Siri analysed the errors attributable to changes in hydration in calculations of body composition from body density measurements. He then derived a three-compartment model in which simultaneous measurements of body weight, density and total-body water allow the division of the body into fat, water, and protein plus mineral compartments. Firstly total-body water is measured and its contribution to total-body density is removed so that the density of the anhydrous body, consisting of fat and protein plus mineral, can be calculated. The proportion of fat is calculated assuming the densities of fat (0.90 kg/l) and protein plus mineral (1.51 kg/l) are known and constant. The overall equation for calculating the proportion of body fat is:

$$f = \frac{2.118}{d - 0.78\,w - 1.354}$$

where f is fat as a fraction of body weight, w is water as a fraction of body weight and d is body density (kg/l).

A modification of the Siri approach has been made by Murgatroyd & Coward (1989) for the measurement of changes in body composition. Often in nutrition research the change in an individual, in response to a change in diet, exercise or disease, is of more importance than measurements of the absolute body composition. In the Murgatroyd & Coward model it is assumed that mineral remains constant over short periods of time. Thus the link between protein and mineral, inherent in the Siri three-compartment model, is broken since in the short term the changes in the protein fraction of fat-free mass are unlikely to be matched by parallel changes in mineral. With measurements of the change in body weight, volume and water content simultaneous equations may be solved to estimate the change in fat and protein mass:

$$\Delta f = 2.741\ \Delta V - 0.7148\ \Delta tbw - 2.046\ \Delta W$$
$$\Delta p = 3.046\ \Delta W - 0.2852\ \Delta tbw - 2.741\ \Delta V$$

where Δf is change in fat mass, Δp is change in protein mass, ΔV is change in volume, Δtbw is change in TBW mass and ΔW is change in weight.

The potential improvement in the estimate of composition change by using this three-compartment model over that of water or density can be calculated (Murgatroyd & Coward, 1989). In a hypothetical 70 kg man who loses 4.2 kg of his body weight, including 0.6 kg of fat, 3.2 kg water and 0.4 kg protein, density measurements overestimate fat loss threefold (– 1.82 kg), total-body water measurements show a net gain of fat

(+ 0.21 kg), whilst the three-compartment model measures the change in both fat and protein accurately.

However, for absolute measurements of fat mass the measurement of bone mineral is extremely valuable. Over the last few years dual-energy X-ray absorptiometry (DXA) has allowed the measurement of bone mineral in a simple and non-invasive manner. In addition it offers the division of soft tissue into fat and the remaining fat-free mass. Thus in itself DXA is a three-compartment model, dividing the body into bone, fat and fat-free soft tissue. However, the accuracy of the soft tissue measurements by DXA alone is still in question (Jebb *et al.*, 1995).

Four-compartment models

DXA may be combined with other techniques in a variety of multi-compartment models. For example, DXA plus total-body water gives mineral, water, fat and the remaining fat-free soft tissue which is predominantly protein. More usefully, if the facilities are available for the measurement of body weight, bone mineral, body density and total-body water, it is possible to derive an alternative four-compartment model including fat, water, protein and mineral. This is the approach which we have employed in many of our recent studies of body composition (Fuller *et al.*, 1992). In this model the contributions of both water and mineral to the total-body density are removed. The remaining compartment can then be divided by assuming the density of fat to be 0.9 kg/l and that of protein to be 1.34 kg/l. Here the errors in assumed densities are small since they each represent a more homogeneous compound than in the previous models. Then, as previously, the proportion of fat may be calculated. This is illustrated in Fig. 13.2.

An alternative four-compartment model has been described by Heymsfield *et al.* (1990) in which *in vivo* neutron activation analysis (IVNAA) is used to measure total-body calcium as an estimate of mineral mass, and nitrogen as an indicator of protein. With the additional measurement of total-body water, body weight may be reconstructed as fat, water, protein and mineral.

There are both theoretical and practical differences in these two approaches. The first is the failure of either of these four-compartment models to account for glycogen. In the former, glycogen will tend to be included in the protein compartment since the density of glycogen (1.5 kg/l) is similar to that of protein (1.34 kg/l), whereas in the IVNAA model glycogen remains unmeasured and so will be included in the difference between the measured components and body weight, i.e. the fat compartment. Nucleic acids (approximately 200 g in a 70 kg man, with a density of

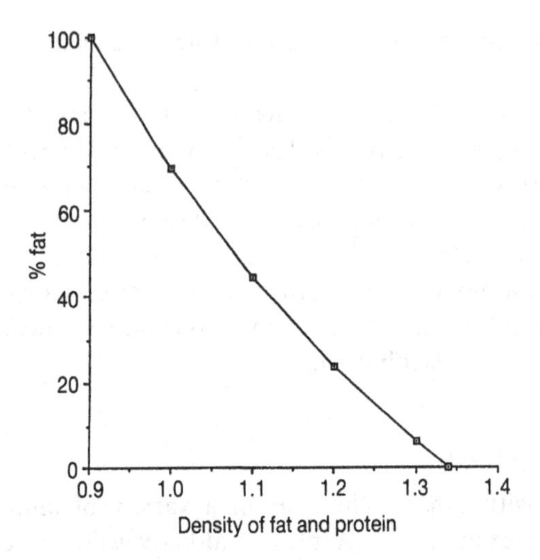

Fig. 13.2. Calculation of body fat from density measurements. Density of fat is assumed to be 0.90 kg/l; density of protein is assumed to be 1.34 kg/l.

1.3 kg/l) and other nitrogenous substances, will tend to be included in the protein fraction with both models. The propagated measurement precision is approximately 1% of body weight for both models.

In practice both approaches require access to a range of body composition techniques which are unlikely to be found outside specialist body composition laboratories. The IVNAA method is significantly more specialised, expensive and exposes the subject to a considerable radiation dose (2.5 mSv). This is in contrast to the body density model where a single whole-body DXA scan for the measurement of mineral mass gives a radiation dose in the order of a day's background radiation (2–5 μSv, depending on the machine). However, the IVNAA model requires considerably less subject co-operation than the model involving underwater weighing, and many volunteers may prefer to be irradiated than face a dip in an underwater weighing tank! Certainly for patients the IVNAA model is often a more realistic measurement option.

Given the current technology a host of alternative variations can be considered. Using IVNAA it is possible to refine the model further by dividing total mineral into osseous and cell mineral. In the six-compartment model Heymsfield & Waki (1991) also include an estimate of glycogen. This is based on the prediction of the glycogen content of the body from the amount of body protein. This, however, is not ideal since there are many situations in which this relationship may be broken, such as overfeeding, starvation, disease and the composition of the diet consumed. Other possibilities include the division of total-body water into intra- and

Table 13.2. *Calculated measurement precision of a variety of body composition methods*

Method	Raw measurement (% CV)	Fat mass (kg)
Density	0.21–0.23	0.73–0.79
Total-body water	1.0–2.0	0.55–1.10
Total-body potassium	2.0–3.0	1.1–4.7
Three-compartment		0.5–0.75
Four-compartment		0.55–0.75

From Elia (1992).
CV, coefficient of variation.

extracellular components using a second isotope dilution, such as bromide, to measure the extracellular fluid volume.

Each refinement of the model increases the complexity of measurement, yet improves the theoretical accuracy with which the body is described, since it makes fewer assumptions about the relationships between compartments. It is always important to ensure that the improved accuracy is not counterbalanced by a decrease in precision caused by measurement errors. Table 13.2 shows the calculated methodological precision for a variety of two-, three- and four-compartment models. It is apparent that the three-compartment method, combining measurements of body density and water, and the four-compartment model, which additionally includes DXA, are more precise than the classical two-compartment models, density, TBW and TBK.

The use of four-compartment models to determine the mass and composition of fat-free mass

There are several studies in the literature in which a four-compartment model has been used to determine the hydration fraction of fat-free tissue and its bone mineral content. In this way it is possible to begin to investigate the error which arises from the assumption that these variables remain constant in the two-compartment model.

Since water is measured independently it is possible to use this approach to estimate the water content of fat-free mass. Data from a number of studies are shown in Fig. 13.3. It is clear that there is great variability between individuals both within and between studies. In our own study (Fuller *et al.*, 1992) of 28 subjects the hydration fraction ranged from 69.4% to 78.4 %, with a mean (± SE) of 73.8% (± 2.13%) and no significant difference between men and women. In the study by Friedl *et al.* (1992) the mean hydration was reported to be 72.6% with a range of 71.1–74.2%. The small variability between individuals in this study may be due to the

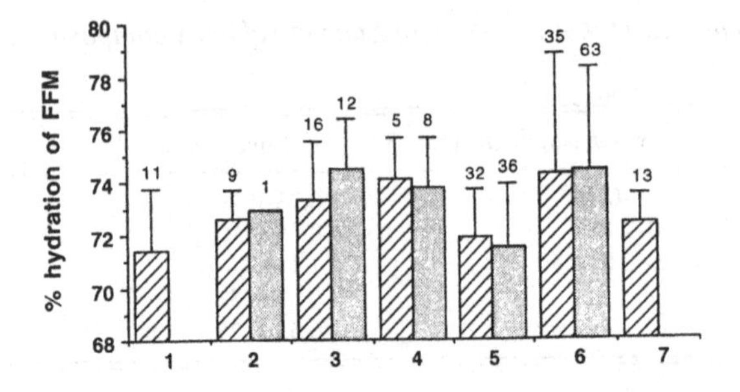

Fig. 13.3. Calculated hydration fraction of fat-free mass (FFM) using a four-compartment model. Hatched columns, male; stippled columns, female. 1, Goran *et al*., 1993 ('young adult'); 2, Friedl *et al*., 1992 (19–24 years); 3, Fuller *et al*., 1992 (18–59 years); 4, Heymsfield *et al*., 1989 (24–94 years); 5, Williams *et al*., 1993 (49–82 years); 6, Baumgartner *et al*., 1991 (>65 years); 7, Goran *et al*., 1993 ('elderly').

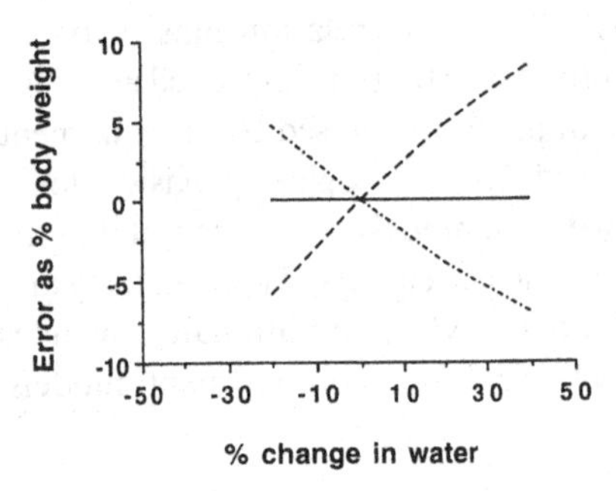

Fig. 13.4. Theoretical errors in the calculated fat (as % body weight) caused by changes in the hydration fraction of fat-free mass: ---, density; -·-·-, total-body water; ——, four-compartment model.

homogeneous group of subjects studied, in terms of age, body mass index, etc., and to the use of the mean of multiple measurements of TBW and density for individual subjects.

It is possible to calculate the effect of changes in the hydration fraction on the calculation of fat mass in Reference Man (Elia, 1992). Fig. 13.4 shows the theoretical errors incurred by a change in the hydration of fat-free mass whilst the other components remain constant. An increase in the water content will cause an overestimation of the fat mass by density yet a decrease in the fat estimate by total-body water measurements. A four-compartment model in which water has been measured independently remains unaffected by any change in hydration.

Since bone mineral is measured independently by DXA it is possible to calculate the mineral fraction of fat-free mass. Fig. 13.5 shows the results

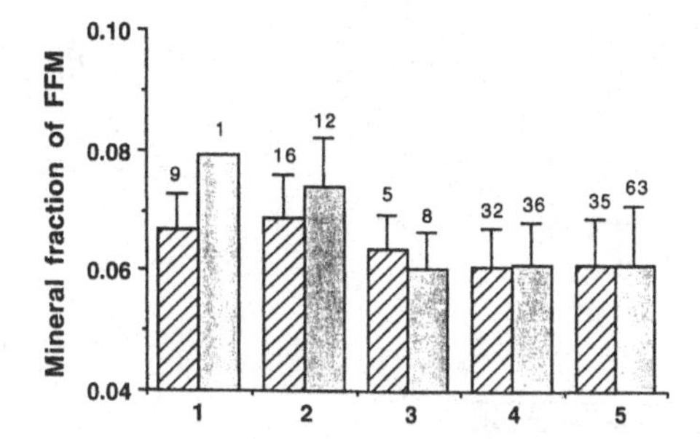

Fig. 13.5. Calculated mineral content of fat-free mass (FFM) using a four-compartment model; Hatched columns, male; stippled columns, female. 1, Friedl *et al.*, 1992 (19–24 years); 2, Fuller *et al.*, 1992 (18–59 years); 3, Heymsfield *et al.*, 1989 (24–94 years); 4, Williams *et al.*, 1993 (49–82 years); 5, Baumgartner *et al.*, 1991 (>65 years).

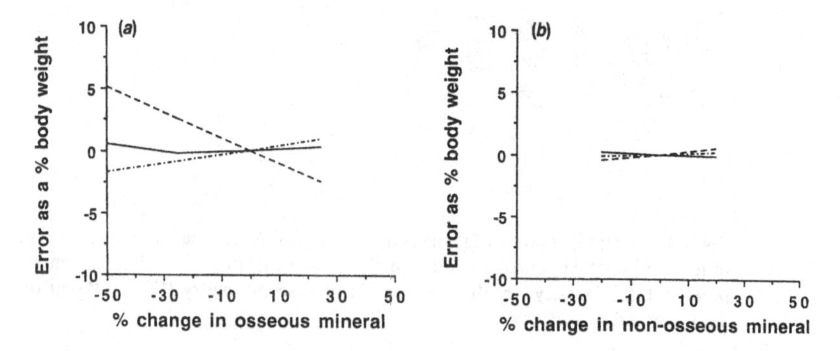

Fig. 13.6. Theoretical errors in the calculated fat (as % body weight) caused by changes in the composition of fat-free mass: ---, density; -·-·-, total-body water; ——, four-compartment model. (*a*) Osseous mineral; (*b*) non-osseous mineral.

from a number of studies, in which the mean mineral content ranges from approximately 6% to 8%. The greater proportion of mineral in the two studies of Friedl *et al.* (1992) and Fuller *et al.* (1992) probably reflects the greater proportion of young, active subjects in these groups.

The calculated theoretical errors caused by changes in bone mineral are shown in Fig. 13.6. A decrease in osseous mineral of 25%, which may occur in severe osteoporosis, will lead to an overestimation of fat mass by 2.5 % by density and an underestimate of almost 1% by total-body water measurements. There are also small errors in the four-compartment model since this assumes a fixed relationship between osseous and non-osseous mineral. Changes in non-osseous mineral cause errors in a similar manner to those for osseous mineral. However, the effect is much smaller since cellular mineral represents only 18% of total mineral mass.

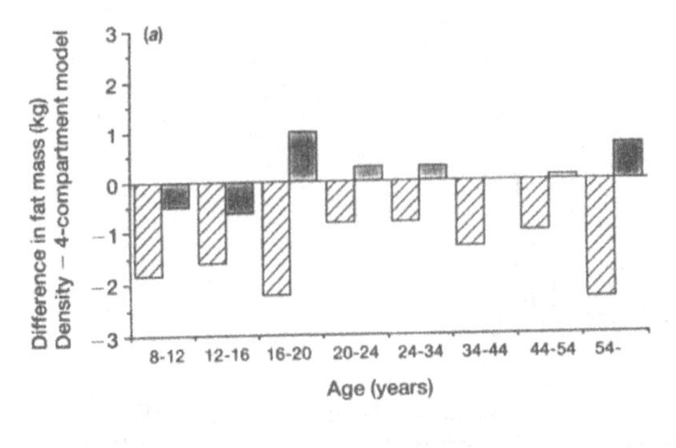

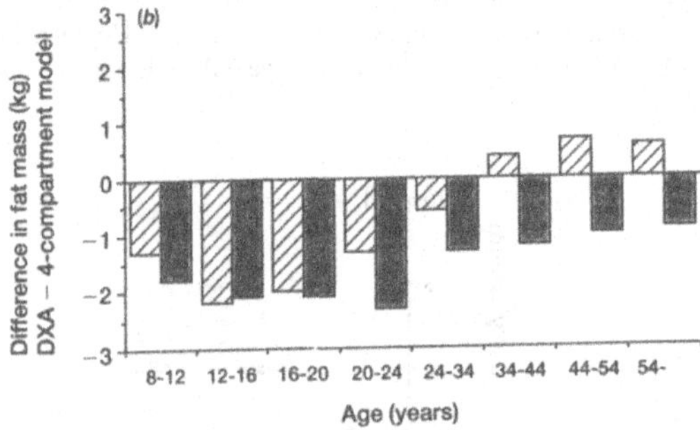

Fig. 13.7. A comparison of fat measured using a four-compartment model and other methods: Hatched columns, male ($n = 100$); stippled columns, female ($n = 125$). (*a*) Density; (*b*) dual-energy X-ray absorptiometry (DXA). Based on data from Guo *et al.* (1992).

Comparisons between two-, three- and four-compartment models

The variation in composition of individual subjects, away from that of Reference Man, explains the differences between classical two-compartment models and the refined four-compartment technique. The largest comparison of four-compartment models against alternative simpler methods in the literature to date seems to be that of Guo *et al.* (1992) who have studied 100 male and 125 female subjects aged from 8 to 68 years as part of the Fels Longitudinal study. Fig. 13.7*a* shows the differences between densitometry and a four-compartment model at different ages. In females the difference is less than 1 kg fat in all groups, whereas in men the error is up to 2 kg, and densitometry consistently yields a lower estimate of body fat mass.

In comparison with DXA (Fig. 13.7*b*) the differences are generally greater in both sexes. For females DXA always gives lower values for fat than a four-compartment model. In males there is a trend for an underestimate of DXA-derived fat compared with a four-compartment

Table 13.3. *Bias and 95% limits of agreement (±2 SD) between measurements of body fat (as % body weight) by a four-compartment model and other methods*

Bias ± 2 SD	Four-compartment model
Three-compartment	1.23 ± 1.43
DXA	1.36 ± 4.95
Density	0.75 ± 3.78
Total-body water	1.70 ± 3.91
Total-body potassium	−0.60 ± 10.42
BMI formula (Black *et al.* 1983)	0.71 ± 8.21
Skinfold thicknesses	0.09 ± 5.42
Resistance (Valhalla equation)	3.39 ± 7.30
Near-infrared interactance	0.41 ± 7.79

From Fuller *et al.* (1992)
DXA, dual-energy X-ray absorptiometry; BMI, body mass index.

model in the younger age groups; this gradually diminishes with age and beyond 34 years becomes a slight overestimate.

In our own studies (Fuller *et al.*, 1992) we have compared results from a four-compartment model with those from two- and three-compartment models and with several bedside techniques. In Table 13.3 the methods have been compared in the manner described by Bland & Altman (1986), in which the mean difference between the two methods has been calculated. This measures the bias of one method relative to the other. In addition the standard deviation of the differences between methods gives an indication of the extent of scatter of individual results. Thus good agreement between methods is demonstrated by a small bias and small confidence limits, which are expressed as ± 2 SD.

Not surprisingly the three- and four-compartment models agree closely since the only difference between them lies in the additional measurement of bone mineral. The error against DXA, density and total-body water is of the order of 1–2 kg fat, with 2 SD representing some 4–5 kg fat. The confidence limits for potassium are large, probably reflecting problems in potassium counting which may be specific to our counter.

The comparison of the body mass index, skinfolds, resistance and near-infrared interactance (NIRI) with a four-compartment model shows that although the mean difference is encouragingly small, the confidence limits are large. This reflects the doubly-indirect nature of the bedside measurements, which are all methods designed to predict the outcome of a two-compartment model. The agreement between skinfold thicknesses and a four-compartment model is better than for other bedside methods in this group of subjects. The results for the resistance technique relate to the equations specific to the Valhalla machine and different results would be obtained with other equations. Overall these bedside methods perform well on a group basis but extremely large errors are possible in individual subjects.

In this way the four-compartment model may be used as an *in vivo* 'gold standard' against which to validate other, simpler body composition techniques. In the past cadaver analysis has been viewed as the only appropriate validation for *in vivo* techniques, but it is not without its own drawbacks and there are obviously severe limitations to the number and type of subjects which may be studied. Loss of water after death can undermine any results, whilst the accurate handling and chemical analysis of such large sample volumes presents numerous difficulties and potential for error. Heymsfield *et al.* (1989) have elegantly demonstrated how body weight may be reconstructed to within 3 % from the summed weights of individually measured components and this is a powerful argument for the acceptance of multi-component models as an *in vivo* gold standard.

Practical applications of four-compartment models

Currently studies of multi-compartment models are confined almost exclusively to methodological considerations, although a few studies are now beginning to appear in which these methods are being applied to measure body composition in different states, both normal and pathological.

Heymsfield *et al.* (1993) have used a six-compartment model to look at age-related changes in body composition. In a group of subjects aged from 26 to 93 years there was no significant change in hydration or the density of fat-free mass. There was a tendency for hydration to increase very slightly with age and for the density of fat-free mass to diminish marginally. However, very significant decreases in the potassium content of fat-free mass were noted. It appears that it is only in young men that the potassium content is close to that typically assumed in the two-compartment model to estimate fat from total-body potassium measurements.

Lonn *et al.* (1992) have used a four-compartment model to measure changes in fat mass before and after 26 weeks of growth hormone administration in children with pituitary insufficiency. The decrease in fat is smaller when measured by a four-compartment model (– 4.2 kg) than by impedance (– 6.0 kg) or computed tomography (– 6.2 kg).

Conclusion

Body composition analysis has now reached a stage where multi-compartment models have been clearly shown to offer improved accuracy, without loss of precision for body composition analysis. By removing many of the assumptions about the relationship of compartments to each other it is possible to account for much more of the great biological diversity seen amongst different populations and in patients whose body composition is distorted by disease. Ultimately it should be possible to develop more definitive equations which will allow improved accuracy in classical two-compartment methods by establishing the true steady-state relationships between different compartments at different ages and in disease.

The development of multi-compartment models has been made possible by great strides in technology, notably DXA and IVNAA. Such methods will in due course allow us to establish reference data for normal populations, to make accurate measurements of changes in individual subjects with disease, when the normal relationships between different compartments may be distorted, and to act as an *in vivo* gold standard for the development and validation of other, simpler methods.

It may also be possible to divide the body in different ways. In this chapter attention has been paid to the measurement of the major substances in the body as a whole. However, the body may also be divided into its elements, the composition of cellular and non-cellular components and the composition of individual body segments. In the past such measurements have not been possible due to the lack of available technology, but today several such measurements are feasible and in this way the marriage of new concepts to new technology in body composition research seems certain to continue.

References

Baumgartner, R.N., Heymsfield, S.B., Lichman, S., Wang, J. & Pierson, R.N. (1991). Body composition in elderly people: effect of criterion estimates on predictive equations. *American Journal of Clinical Nutrition*, **53**, 1345–53

Black, D., James, W.P.T. & Besser, G.M *et al.* (1983). Obesity. A report of the Royal College of Physicians. *Journal of the Royal College of Physicians*, **17**, 5–65.

Bland, J.M. & Altman, D.G. (1986). Statistical methods for assessing agreement between two methods of clinical measurement. *Lancet*, **i**, 307–10.

Elia, M. (1992). Body composition analysis: an evaluation of 2 component models, multicomponent models and bedside techniques. *Clinical Nutrition*, **11**, 114–27.

Forbes, G.B., Gallup, J. & Hursh, J.B. (1961). Estimation of total body fat from potassium-40 content. *Science*, **133**, 101–2.

Friedl, K.E., de Luca, J.P., Marchitelli, L.J. & Vogel, J.A. (1992). Reliability of body fat estimations from a four-compartment model by using density, body water and bone mineral measurements. *American Journal of Clinical Nutrition*, **55**, 764–70.

Fuller NJ, Jebb SA, Laskey MA, Coward WA, Elia M. (1992). Four-component model for the assessment of body composition in humans: comparison with alternative methods and evaluation of the density and hydration of fat-free mass. *Clinical Science*, **82**, 687–93.

Goran, M.I., Engelberg, T.E., Poehlman, E.T., Nair, S. & Danforth, E. (1992). Comparison of body composition derived by density, water and a 3-compartment model combining density and water. Presented at the *in vivo* body composition symposium 1992. Houston, USA.

Guo, S.S., Chumlea, W.C., Wu, X., Wellens, R., Siervogel, R.M. & Roche, A.F. (1992). A comparison of body composition models. Presented at the *in vivo* body composition symposium 1992. Houston, USA.

Heymsfield, S.B. & Waki, M. (1991). Body composition in humans: advances in the development of multicompartment chemical models. *Nutrition Reviews*, **49**, 97–108.

Heymsfield, S.B., Wang, J., Kehayias, J., Heshka, S., Lichtman, S. & Pierson RN. (1989). Chemical determination of human body density *in vivo*: relevance to hydrodensitometry. *American Journal of Clinical Nutrition*, **50**, 1282–9.

Heymsfield, S.B., Lichtman, S., Baumgartner, R.N. *et al.*, (1990). Body composition of humans: comparison of two improved four-compartment models that differ in expense, technical complexity and radiation exposure. *American Journal of Clinical Nutrition*, **52**, 52–8.

Heymsfield, S.B., Wang, Z., Baumgartner, R.N., Dilmanian, F.A., Ma, R. & Yasumura, S. (1993). Body composition and aging: A study by *in vivo* neutron activation analysis. *Journal of Nutrition*, **123**, 432–7.

Jebb, S.A., Goldberg, G.R., Jennings, G. & Elia, M. (1995). Dual energy x-ray absorptiometry measurements of body composition: effects of depth and tissue thickness, including comparisons with direct analysis. *Clinical Science* (in press).

Lonn, L., Bengtsson, B.A., Bosaeus, I., Kvist, H., Tolli, J., & Sjostrom, L. (1992). Growth hormone induced changes in body composition in subjects with pituitary insufficiency. Presented at the *in vivo* body composition symposium 1992. Houston, USA.

Murgatroyd, P.R. & Coward, W.A. (1989). An improved method for estimating changes in whole-body fat and protein mass in man. *British Journal of Nutrition*, **62**, 311–14.

Siri, W.S. (1961). Body composition from fluid spaces and density: analysis of methods. In: *Techniques for measuring body composition*, ed. J. Brozek & H. Henschel. Washington, D.C.: National Academy of Sciences.

Williams, D.P., Going, S.B., Massett, M.P., Lohman, T.G., Bare, L.A. & Hewitt, M.J. (1992). Aqueous and mineral fractions of the fat-free body and their relation to body fat estimates in men and women aged 49–82 years. Presented at the *in vivo* body composition symposium 1992. Houston, USA.

14 *The future of body composition research*

STEVEN B. HEYMSFIELD AND ZI-MIAN WANG

Introduction

Interest in the study of human body composition spans at least a hundred years. The importance of body composition research as a distinct science is evident during at least two main periods during this century. The first period occurred immediately following World War II with the investigation of human semi-starvation and proceeded to early studies of body composition in disease. Many techniques for evaluating body composition were developed during this period, particularly tracer methods for quantifying specific components such as total-body water and fluid volumes. The excellent research during this period is summarised in the first body composition symposium of the Society for the Study of Human Biology held in 1963 and published in 1965 (Brozek, 1965).

The second major era in body composition research began during the mid-1980s with the recognition that many chronic and acute illnesses involved alterations in body composition, that these changes were linked with morbidity and mortality, and that nutritional treatments could affect patient outcome. Many body composition techniques were either introduced or fully appreciated during this era, which we are still in, including important new imaging, neutron activation analysis, and electrical body composition methods (Forbes, 1987).

What makes the present era unique is the recognition by some investigators that the study of body composition *per se* is a distinct area of human biology. There are now many investigators who spend the majority of their time studying various aspects of human body composition; many research reports are appearing in scientific journals; and there are research symposia at increasingly frequent intervals. It is this recognition that body composition is emerging as a mature area of scientific investigation that stimulates us not only to reflect on past accomplishments, but to also look ahead to future developments. A first step in speculating on future developments in body composition research is to define the boundaries of the field as it now exists.

The study of human body composition

Several years ago research reports describing body composition methodology and body composition changes in various conditions were appearing at a rapid rate. A few excellent textbooks on body composition were also published (Forbes, 1987; Lohman *et al.*, 1988; Himes, 1991; Lohman, 1992). It thus seemed as though there was a 'critical mass' of information emerging in the field, although there appeared not to be any coherent scheme for organising this knowledge. Different and often ill-defined terms were used for the same body composition component. Methods were presented by author preference and there appeared to be no logical manner in which students and new investigators could build an understanding of the field grounded on a strong scientific foundation. It thus appeared to us that a first step in securing a strong future for the study of human body composition was to establish a clear organisation of the field as it exists today. Additionally, it was our hope that in the process of organising body composition information we would identify previously unforeseen research opportunities.

Our initial impression was that there are two main distinct areas of body composition research: methodology and biology (Wang *et al.*, 1992). However, there appeared to be no prior attempt at systematically organising the human body into distinct compartments or components. This missing information was evident first when we tried to assemble a listing of the various components as defined by other investigators. Confusing and overlapping terms abounded: lean-body mass, fat-free body mass and adipose-tissue-free mass, for example, were often used synonymously and yet each one lacked a clear definition. As a result of this confusion investigators often erroneously developed models in which components at the same body composition level overlapped with each other and thus when combined did not provide a true representation of body weight. This observation led us to appreciate the existence of a third and as yet unrecognised area of body composition study: the five-level model and its associated rules (Wang *et al.*, 1992). The five-level model (Fig. 14.1) of body composition represents the classic organisational levels of biology: atomic, molecular, cellular, tissue-system and whole-body. Our formulation of body composition study thus included three interrelated areas (Fig. 14.2): the five-level model and its associated rules, methodology and biological effects. Each area represents the science, technology and application of body composition investigation. All present body composition information can be organised into these three areas.

Five-level model and the future

The five-level model was developed by us in an attempt to clarify the definition of various components and to aid in examining the relations between components. Each component at the five levels of body composition,

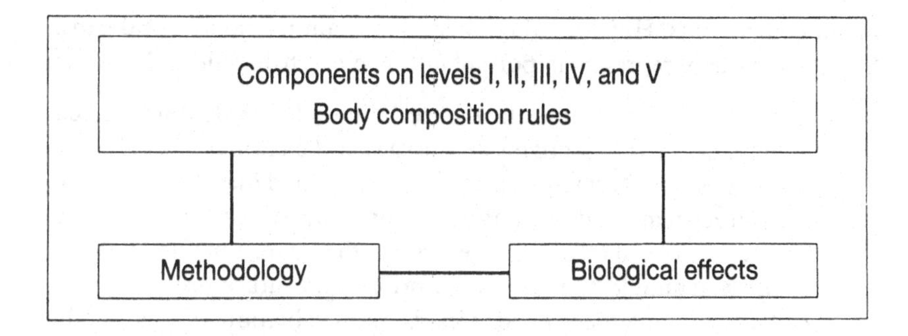

Fig. 14.1. The three areas that comprise the study of human body composition. From Wang *et al.* (1992) with permission.

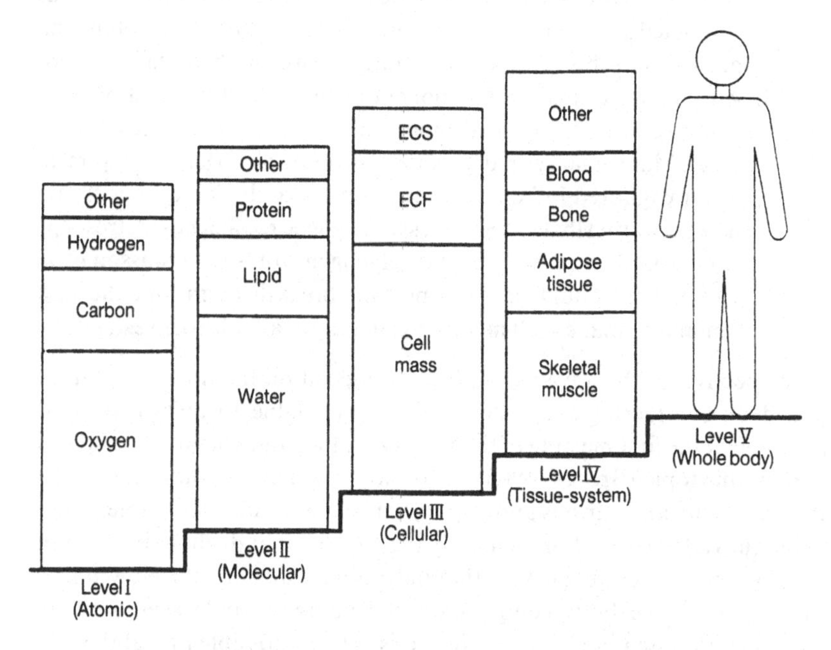

Fig. 14.2. The five-level model of human body composition. ECF and ECS, extracellular fluid and solids, respectively. From Wang *et al.* (1992) with permission.

shown in simplified form in Fig. 14.2, exists at a specific level, has a precise definition, and does not overlap with other components at the same level. A complete description of the five-level model of body composition is provided in Wang *et al.* (1992).

Once we had a clear picture of human body composition, it became apparent that methods were available for estimating most of the 30 or more major components. However, there were important methodological gaps. Several of the gaps represent important future research directions. Notably:

1. There is a conspicuous lack of techniques for evaluating the cell mass of individual organs and tissues at the cellular level of body composition. Numerous presently unanswered questions could be explored if methods were available for quantifying the mass of each category of cells. This area represents the clearest methodological gap and should thus be a high priority for future research.
2. Another underdeveloped area is the evaluation of total-body skeletal muscle, a component at the tissue-system level of body composition. This is surprising, as skeletal muscle is one of the largest components of body weight and so much interest exists in skeletal muscle mass and function. Imaging techniques such as computed axial tomography show promise as methods for estimating muscle mass, but substantially more research in this area is needed.
3. Although the smallest component at the molecular level of body composition, glycogen is important to several different research areas. Many unanswered questions surround glycogen, a component usually ignored in body composition research. Initial reports are now appearing that suggest nuclear magnetic resonance spectroscopy can be used to quantify hepatic glycogen stores (Magnusson *et al.* 1992). This would be an important breakthrough, but the task remains to make such methods generally available to investigators.

Perspective on the advances in body composition methodology can be gained by considering what methods were available to investigators 30 years ago at the first meeting of the Society for the Study of Human Biology held on this topic (Brozek, 1965). Underwater weighing, total-body water and potassium and various anthropometric methods were available. These techniques allowed analysis of mainly the molecular and whole-body levels of body composition. Even then, the analyses were incomplete. Now almost all components of body composition can be analysed *in vivo*, with the exception of those noted above. Thus, it is entirely possible that at the next Society meeting all major components of body weight will be measurable *in vivo*.

Another important category of research also became clear once we had developed the five-level model. We recognised that connections existed between components at the same or different levels. For example, fatness can be described at levels I (atomic) to V (whole-body) in consecutive order as total-body carbon, lipid or fat, adipocytes, adipose tissue and skinfolds. Each is a different manifestation of fatness although all are linked by common factors. We term these factors steady-state relations and they can

Table 14.1. *Steady-state relations in 19 matched young and old Caucasian females*

	Young	Old	*p*
Age (years)	29.9 ± 4.4	74.2 ± 6.7	0.001
Weight (kg)	59.6 ± 6.6	59.7 ± 7.2	NS
Height (m)	1.63 ± 0.05	1.62 ± 0.05	NS
TBW/FFM (kg/kg)	0.735 ± 0.020	0.725 ± 0.030	NS
TBK/FFM (mmol/kg)	59.8 ± 4.4	54.1 ± 3.4	0.001
FFM density (g/cm^3)	1.102 ± 0.007	1.101 ± 0.008	NS

Modified from Mazariegos *et al.* (1994).
Values are the mean ± SD.
FFM, fat-free body mass; TBK and TBW, total-body potassium and water, respectively.

be described by mathematical equations. These steady-state relations are not only conceptually important, but play a central role in body composition methodology.

At present our knowledge of relations between components is rudimentary. For example, in adults we assume that some components are related to each other in a stable and predictable manner independent of gender, age or ethnicity (Heymsfield & Waki, 1991). This is a vital assumption, as will be shown in the next section, because many body composition methods employ these steady-state relations by assuming that a stable and known relation exists between the components (Heymsfield & Waki, 1991). Most of these steady-state relations were developed at a time when the only way to evaluate many components was by direct analysis of cadavers (Brozek *et al.*, 1963). Now it is possible to examine almost all components at the five levels of body composition *in vivo* and to establish steady-state relations for specific groups. A few examples are the following:

1. It is generally assumed that young and old subjects have similar steady-state relations between the components of fat-free body mass (FFM), a molecular level compartment. Three of these assumed relations are shown in Table 14.1: the hydration, potassium content, and density of FFM (total-body water/FFM, total-body potassium/FFM, and combined densities of water, protein and mineral) (Mazariegos *et al.*, 1994). We examined these relations using new multi-compartment techniques in height- and weight-matched young and old Caucasian females (Mazariegos *et al.*, 1994). Our results indicated similar hydration and density of FFM between young and old subjects, although the elderly females had a significantly reduced total-body potassium content of FFM. This study is presented as an example of how steady-state relations

Table 14.2. *Steady-state relations in 19 matched Caucasion and Black females*

	Caucasian	Black	*p*
Age (years)	43.7±15.6	44.4±15.4	NS
Weight (kg)	62.0±6.6	62.3±5.8	NS
Height (m)	1.62±0.06	1.61±0.06	NS
TBW/FFM (kg/kg)	0.747±0.037	0.734±0.041	NS
TBK/FFM (mmol/kg)	56.6±7.4	63.1±8.4	<0.03
FFM density (g/cm^3)	1.097±0.009	1.106±0.013	<0.05

Modified from Ortiz *et al.* (1992).
Values are the mean±SD.
FFM, fat-free body mass; TBK and TBW, total-body potassium and water, respectively.

Table 14.3. *Comparison of body composition components in 24 matched Caucasian and Black males*

	Caucasian	Black	*p*
Age (years)	39.7±10.3	39.8±10.3	NS
Weight (kg)	78.0±10.0	78.3±10.6	NS
Height (m)	1.77±0.07	1.77±0.07	NS
TBW (kg)[a]	48.3±5.9	49.3±6.4	NS
TBK (mmol)	3981±547	3900±526	NS
R (ohms)	446±39	486±66	<0.05

Modified from Gerace *et al.* (1994).
Values are the mean±SD.
R, total-body resistance; TBK and TBW, total-body potassium and water, respectively.
[a] Tritium dilution volume.

may either stay the same or change with age. Additionally, this study demonstrates that it is possible to study these steady-state relations *in vivo*.

2. Steady-state relations are assumed to be independent of ethnicity. Tables 14.2 and 14.3 examine this hypothesis in age-, height- and weight-matched Caucasian and Black females and males, respectively. We observed a significantly greater total-body potassium content and density of FFM in the Black females compared with their Caucasian matched counterparts, while hydration of FFM was similar between the two groups (Table 14.2) (Ortiz *et al.*, 1992). The Caucasian and Black males had similar total-body water and potassium, while they differed significantly in total-body resistance (Table 14.3) (Gerace *et al.*, 1994).

The main point of these examples is that characterisation of the relations between body composition components is now possible and is an important

Property	property-based methods →	Cp	component-based methods →	Cu
γ-rays (1.40 MeV)	^{40}K counting	TBK		FFM, fat, BCM
γ-rays (10.8 MeV)	NAA	TBN		protein, sulphur
γ-rays (3.10 MeV)	NAA	TBCa		Mo, ECS
β-rays	$^{3}H_2O$ dilution	water		FFM, fat
γ-, β-, rays	^{22}Na, ^{42}K dilution	Na_e, K_e		ECF, ICF, water
γ-rays	NAA	TBN, TBK		SM
electrical conductivity	TOBEC	total body fluid		FFM, fat
electrical impedance	BIA	total body fluid, ECF		FFM, fat
body volume	UWW	fat, FFM		water

Fig. 14.3. Some of the connections between property-based and component-based body composition methods. Properties are used to estimate about one-half of the components (C_p) at the five levels of body composition. The remaining components (C_u) are calculated from C_p and are referred to as component-based methods. BCM, body cell mass; BIA, bioimpedance analysis; e, exchangeable; ECF, extracellular fluid; FFM, fat-free body mass; ICF, intracellular fluid; Mo, osseous mineral; NAA, neutron activation analysis; TB, total-body; TOBEC, total-body electrical conductivity; UWW, underwater weighing.

area for future research. The relevance of this effort is two-fold: many simple methods rely on these associations and exploring them improves or complements our understanding of many other areas of human biology and physiology.

Methodology and the future

Body composition methods are usually presented in the chronological order of their development or in groups related to specific components such as fat, skeletal muscle, elements and fluid (Forbes, 1987). These traditional approaches to organisation do not in themselves reveal new information about the methods or suggest future possibilities. We are now developing an organisation plan for body composition methods in an effort to examine future methodological possibilities.

At present all body composition methods can be organised into two main groups according to the relation (Fig. 14.3):

$$\text{Property} \xrightarrow{\text{property-based methods}} \text{Component 1 (C1)} \xrightarrow{\text{component-based methods}} \text{Component 2 (C2)}$$

The first group of methods (left-hand portion of Fig. 14.3), which are used to estimate about one half of the 30 major body composition components, are referred to as property-based methods. This is because the component (C1) is calculated from a measured property using a regression equation (e.g. total-body resistance for calculating total-body water using regression equations derived on reference groups) or a ratio equation (e.g. body density for calculating total-body fat using the Siri model) (Siri, 1956). The second group of methods (right-hand portion of Fig. 14.3), which are used to estimate the remaining components, are referred to as component-based as they have in common the estimation of component 2 (the unknown) from component 1. C2 can be calculated from C1 using a regression equation (e.g. total-body water to calculate total-body protein) (Beddoe *et al.*, 1985) or a ratio equation (e.g. total-body water to calculate FFM). This approach to classifying methods is expanded upon in a more detailed report (Wang *et al.*, 1995). In the context of this overview, can our simple methodological organisation suggest future research directions?

In providing an answer to this question, we use as an example one specific group of component-based methods. These methods all have in common the use of a ratio (R) as the 'model' from which C2 is calculated from C1. In other words,

$$\mathrm{C1} = \mathrm{R} \times \mathrm{C2} \text{ or } \mathrm{C2} = \mathrm{C1}/R$$

An example is the total-body water (TBW) method of estimating fat-free body mass (FFM) as

$$\mathrm{TBW} = 0.73 \times \mathrm{FFM} \text{ or } \mathrm{FFM} = \mathrm{TBW}/0.73$$

The ratio used in this method (TBW/FFM of 0.73) is based mainly on the steady-state value for the average water content of defatted cadavers (Brozek *et al.*, 1963; Sheng & Huggins, 1979). There are many component-based ratio methods used in body composition research.

While examining these ratios we noted that some were accepted as extremely stable by investigators and others were not. For example, many authors equate total-body nitrogen, which can be measured *in vivo*, very strongly with total-body protein (an unmeasurable component). It appeared to us that nitrogen and protein were indeed very closely linked with each other because of a stable chemical combination (N is 16% protein) and because over 98% of total-body nitrogen is in the form of protein (Snyder *et al.*, 1975). This and other similar observations led us to develop a new concept related to ratio methods, to organise relationships between components into eight groups and to uncover some possible future research areas.

First, we reasoned that steady-state ratios between components were not really constants, but rather averages that represented a distribution of ratios. That is, R is actually R $\pm$ SD (Wang *et al.*, 1995). We have simplified this relation by expressing SD as coefficient of variation (CV) and termed the ratio's CV its 'stability'. In this context stability refers to the biological variability of a ratio and does not consider measurement error. We then compared the ratio $\pm$ CV of some pairs of components. For example,

$$\text{TBW/FFM (kg/kg)} = 0.73 \pm 1\%$$
$$\text{TBN/protein (kg/kg)} = 0.16 \pm \sim 0$$
$$\text{TBK/FFM (mmol/kg)} = 68.2 \pm 4.5\%$$
$$\text{TBC/fat (kg/kg)} = 1.52 \pm 25\%$$

where TBN, TBK and TBC are total-body nitrogen, potassium, and carbon, respectively. These varying stabilities suggested to us that some ratios are much more stable than others and there might be an explanation for these differences.

Our first approach was to group methods according to their combination characteristics: chemical (e.g. nitrogen and protein) or physiological (e.g. TBW and FFM). We also found that another combination between components was neither chemical nor physiological, and we refer to it as 'variable' (e.g. fat with body weight). We found that some chemical combinations had high stability and variable combinations the lowest stabilities. However, there was considerable overlap between the three groups in stabilities. It was then that another set of relations between two components became apparent. We term these 'dependency relations', such that a relation is subordinate if one component is completely (i.e. $>98\%$) within another (e.g. TBW and FFM), overlapping if one component is partially within another (e.g. TBC and fat) and separate if there is no overlap between components (e.g. nitrogen and sulphur). Each ratio and its stability could then be classified into one of eight combination characteristic-dependency groups.

When we organised all relationships between components into these eight groups we found that only three gave high stability ($CV < 10\%$; i.e. chemical subordinate, chemical separate and physiological subordinate) and were ideal for use in monoratio methods. Three had low stability ($CV > 10\%$; i.e. physiological separate, variable subordinate and variable separate). Low-stability relations between components are unsuitable for use in developing monoratio component-based methods and at best might provide a rough estimate of an unknown component.

The accuracy of a component-based ratio method depends mainly upon the stability of the ratio between unknown and known component. This attempt at organising body composition methods serves as an example of how more can be learnt about presently-used methods and also suggests future pathways for new model development. An example of new model development is provided by examination of the eight relationship types. Three of the types appeared to give high stabilities and thus are ideal for developing monoratio methods. For the other two types (i.e. chemical and physiological overlapping), we found that the stabilities are improved if they are developed into multiratio methods. For example, total-body carbon can be used to calculate total-body fat (a chemical overlapping relation) if the model is expanded to include nitrogen to estimate carbon in protein and glycogen and calcium to estimate carbon in bone mineral (Kehayias *et al.*, 1991). Another example is that total-body skeletal muscle is often calculated from 24 h urinary creatinine by assuming that 17 kg of skeletal muscle produces 1 g of urinary creatinine (Heymsfield *et al.*, 1983). This relation between creatinine excretion and skeletal muscle has until recently been considered highly stable. However, in a recent study we found that the ratio of 24 h urinary creatinine to skeletal muscle mass measured by computed axial tomography in healthy men on a meat-free diet was highly variable (i.e. there was a low stability). Further examination of the data revealed that 24 h urinary creatinine plotted against skeletal muscle mass had a non-zero intercept. Our present hypothesis, suggested also by the work of other investigators (Crim, 1976), is that urinary creatinine is also produced in non-skeletal muscle tissues. If this is the case, then a new multiratio equation should be developed to estimate skeletal muscle from creatinine excretion and other measured components. These examples are provided to suggest that there are many possible future new body composition methods that can be created or older ones refined.

Another area within methodology that needs development is the addition of reference bodies beyond that of the 70 kg Reference Man (Snyder *et al.*, 1975). The Reference Man data are an invaluable resource for investigators in all areas of human biological research. The Reference Man represents data assembled from human cadavers and autopsy studies. However, nearly all compartments and components of body weight can now be measured *in vivo*. It thus appears that centres with the required techniques would make a valuable contribution if they developed reference data for specific age, gender and ethnic groups.

Biological effects and the future

Biology is the third and final area in the study of human body composition. We suggest that there are three possible future areas of research in the biological category that relate to the study of body composition as a whole.

First, as with the other areas, there has been little attempt at organising biological information related to body composition. Authors usually present data specific to the methods they employ in their laboratory. Our impression is that a large majority of the body composition research reported in humans describes two main compartments: fat and FFM. Our suggestion is that an investigator interested in a specific human study should first organise all the available information in that area according to the five-level model. This approach will then reveal the major gaps in knowledge in that field. Gradually filling in this information will provide a full understanding of body composition in health and disease.

The second topic in the biological area relates to important clinical and research issues that at present cannot be fully evaluated because of methodological limitations. A few clinical questions in need of new or improved methodology are the following:

Accurate techniques for evaluating short-term changes in body composition, such as with weight loss or gain over several weeks.

Reliable methods of assessing composition of weight gain during pregnancy.

Techniques that can estimate a subject's extra- and intracellular fluid volumes and that can detect small changes in these volumes in the clinical setting.

Methods of evaluating metabolically active tissue that can be used to adjust energy expenditure measurements for between-individual differences in body composition.

Practical approaches to quantifying visceral adipose tissue. This compartment is an important predictor of obesity risk and at present is difficult to measure without excess radiation risk or inordinate expense.

As stated earlier, reliable methods are needed for estimating skeletal muscle mass that can be used in conditions such as wasting illnesses and for evaluating exercise training programmes.

The third potential area of future research in the biological category involves studying the relations between body components and their associated functions. Body composition studies often examine isolated components and fail to establish more satisfying links with various functions at the molecular, cellular, tissue-system and whole-body levels. A few examples are the relations between: skeletal muscle mass, skeletal muscle fibre type and strength or maximal oxygen uptake; heart weight and morphology, myocardial oxygen consumption, and blood pressure; and total-body cellular protein, rate of wound repair, and outcome following major surgical procedures.

This concept relating composition and function could be expanded to the development of dynamic models such as those relating composition of weight change to positive or negative energy-nitrogen balance. Similar models could be developed that relate fluid and electrolyte intake, kidney mass and renal function, and fluid balance. These examples are not intended to be specific, but rather to convey the importance of incorporating body composition information into the broader context of human biology and physiology.

In making those connections between body composition and function, emphasis should be placed on developing models that reflect underlying biochemical and physiological relations. At present many studies, sometimes by necessity, make composition–function associations by empirical means. A contemporary example is resting metabolic rate (RMR), a property that holds considerable interest in many different research areas. Early investigators attempted to explain between-individual (or even between-organism) differences in RMR by differences in body weight, body weight adjusted to various powers, and body surface area (Lusk, 1928). The introduction of body composition techniques in the past several decades led to further studies of the relation between RMR and metabolically active tissue mass (Elia, 1992). Since the main available techniques until recently were those that gave two compartments at the molecular level (i.e. fat and FFM), most studies examine the associations between RMR and FFM. Generally, RMR and FFM are highly correlated in healthy adults (Ravussin & Bogardus, 1989). However, additional significant covariates are often found such as fat mass, age, sex and genetic background (Ravussin & Bogardus, 1989). These additional determinants of RMR are frequently ascribed physiological importance. For example, it is suggested that RMR is reduced in elderly subjects because RMR is lower in old individuals after adjustment for FFM and fat (Ravussin & Bogardus, 1989). A search is under way for the physiological-hormonal basis of this senescence-related decrease in energy expenditure.

While these are important efforts, it should be noted that the associations between RMR and body composition are presently empirical; that is, they are based largely on best-fit regression analyses and not on actual mechanisms. For example, the real association between RMR and body composition at the molecular level is a very complex one that involves the combined energy costs of protein synthesis and breakdown, and those of a myriad of other chemical components. At the tissue-system level, RMR is the net result of energy expended by each organ and tissue. Similar relations between thermogenesis and body composition apply at the cellular level. We need to develop techniques and modelling approaches that give more comprehensive analyses of how composition and function relate to each other at all levels of body composition.

Table 14.4. *Relation between fat-free body mass and fluid components in 19 matched young and old Caucasian females*

Components	Young	Old	*p*
Age (years)	29.9±4.4	74.2±6.7	0.001
TBW (kg)	31.9±4.1	28.7±2.9	0.01
ECF (1)	14.5±2.0	15.6±2.8	NS
ICF (1)	16.1±2.1	13.2±1.4	0.001
BCM (kg)	21.5±2.8	17.8±2.0	0.001
FFM (kg)	43.3±5.2	39.6±3.8	0.001
ECF/ICF	0.92±0.13	1.18±0.18	0.001
BCM/FFM	0.50±0.04	0.45±0.03	0.001

Modified from Mazariegos *et al.* (1994).
Values are the mean±SD.
BCM, body-cell mass; ECF, extracellular fluid; FFM, fat-free body mass; ICF, intracellular fluid; TBW, total-body water.

The present empirical approach, while useful in some contexts, can also lead us astray and thereby devalue the role of body composition research. An example is the lowering of RMR with age after adjustment for FFM using multiple regression analysis. Does this finding imply that a major research effort should be focused on why metabolism slows in the elderly? Or is the finding due to the component chosen as representative of metabolically active tissue? We explored these questions further by evaluating RMR, body composition and ageing in three separate studies of healthy women. In the first study we completed a meta-analysis that indicated RMR decreased significantly with age, although the effect was very small (2% of total variance) after adjustment for body composition (FFM) (Allison *et al.*, 1992). In the second study we examined RMR and body composition in a healthy cohort of 27 White women between the ages of 20 and 86 years. We found that RMR did decrease slightly with age after adjustment for FFM (Buhl *et al.*, 1992). Body cell mass, a cellular level compartment, was also estimated from total-body potassium in the same subjects. We found that RMR increased with age after adjustment for body cell mass, a finding opposite to that for RMR versus FFM. In the third study (Table 14.4) we examined the relations between FFM, extracellular fluid, and body cell mass in healthy young and old matched Caucasian females (Mazariegos *et al.*, 1994). As shown in the table, the old females had a significantly reduced FFM and body cell mass. However, the old females' relative decrease in BCM exceeded that of FFM and the ratio of body cell mass to FFM was significantly lower in the old females. FFM in old females thus has less metabolically active tissue and a relatively larger extracellular fluid compartment than it does in young females. Resting energy expenditure may thus not be lower in old females relative to metabolically active tissue, although much more research is needed at all levels of body composition to develop a comprehensive view of how thermogenesis and specific compartments are related.

Conclusion

Finally, what about the people who do body composition research? A critical element in the field of body composition research is its multidisciplinary nature. Nearly every basic science and clinical field is represented in one way or another. This diversity of the field is one of its great attractions and strengths. On the other hand, it has the potential of leaving some investigators with an allegiance to their primary discipline somewhat isolated. More frequent symposia and even a dedicated journal would help in the future to sustain such a multidisciplinary field.

In conclusion, our review indicates that future research possibilities exist in all three body composition areas: rules that relate components at the five levels of body composition, methodology and biological effects. Few research disciplines offer so many and such diverse opportunities as the study of human body composition.

Acknowledgement

This work was supported by National Institutes of Health Grant PO1-DK42618.

References

Allison, D.B., Heshka, S. & Heymsfield, S.B. (1992). The independent effect of age on metabolic rate: a 'mini-meta analysis'. *American Journal of Clinical Research*, **40**, 2.

Beddoe, A.H., Streat, S.J., Hill., G.L. & Knight, G.S. (1985). New approach to clinical assessment of lean body tissues. In: *Body Composition Assessment in Youth and Adults*, ed. A.F. Roche, pp. 61–71. Columbus, Ohio: Ross Laboratories.

Brozek, J. (1965). *Human Body Composition: Approaches and Applications*. Oxford: Pergamon Press.

Brozek, J., Grande, F., Anderson, T. & Keys, A. (1963). Densitometric analysis of body composition: revision of some assumptions. *Annals of the New York Academy of Science*, **110**, 113–40.

Buhl, K., Lichtman, S., Massucci, L., Wang, J. & Heymsfield, S.B. (1992). Changes in body composition and energy expenditure with ageing in women. *FASEB Journal*, **6**, A1656.

Crim, M. (1976). Creatinine metabolism in men: Creatine pool size and turnover in relation to creatine intake. *Journal of Nutrition*, **106**, 371–81.

Elia, M. (1992). Energy expenditure in the whole body. In: *Energy Metabolism: Tissue Determinants and Cellular Corollaries*, ed. J.M. Kinney & H.N. Tucker. New York: Raven Press.

Forbes, G. (1987). *Human Body Composition in Growth, Aging, Nutrition and Activity*. New York: Springer-Verlag.

Gerace, L., Aliprantis, A., Russell, M. *et al.*, (1994). Differences in skeletal muscle and bone mineral mass between black and white males and their relevance to estimates of body composition. *American Journal of Human Biology*, **6**, 255–62.

Heymsfield, S.B. & Waki, M. (1991). Body composition in humans: advances in the development of multicompartment chemical models. *Nutrition Reviews*, **49**, 97–108.

Heymsfield, S.B., Arteaga, C., McManus, C. & Smith, J. (1983). Measurement of muscle mass in humans: validity of the 24 hour urinary creatinine method. *Americal Journal of Clinical Nutrition*, **37**, 478–94.

Heymsfield, S.B., Lichtman, S., Baumgartner, R.N., *et al.* (1990). Human body composition: comparison of two improved four-compartment models that differ in expense, technical complexity and radiation exposure. *American Journal of Clinical Nutrition*, **52**, 52–8.

Himes, J.H. ed. (1991). *Anthropometric Assessment of Nutritional Assessment of Nutritional Status*. New York: Wiley-Liss.

Kehayias, J.J., Heymsfield, S.B., LoMonte, A.F., Wang, J. & Pierson, R.N. Jr. (1991). *In vivo* determination of body fat by measuring total-body carbon. *American Journal of Clinical Nutrition*, **53**, 1339–44.

Lohman, T.G. (1992). *Advances in Body Composition Assessment*. Champaign, Ill.: Human Kinetics.

Lohman, T.G., Roche, A.F. & Martorell, R., ed. (1988). *Anthropometric Standardisation Reference Manual*. Champaign, Ill.: Human Kinetics.

Lusk, G. (1928). *The Elements of the Science of Nutrition*. Philadelphia: Saunders.

Magnusson, I., Rothman, D.L., Katz, L.D., Shulman, R.G. & Shulman, G.I. (1992). Increased rate or gluconeogenesis in Type II diabetes mellitus: A ^{13}C nuclear magnetic resonance study. *Journal of Clinical Invetigations*, **90**, 1323–7.

Mazariegos, M., Wang, Z., Gallagher, D. *et al.*, (1994). Aging in females: changes in the five levels of body composition and their relevance to the two-compartment chemical model. *Journal of Gerontology*, **49**, 201–8.

Ortiz, O., Russell, M., Daley, T.L. *et al.*, (1992). Differences in skeletal muscle and bone mineral mass between black and white females and their relevance to estimates of body composition. *American Journal of Clinical Nutrition*, **55**, 8–13.

Ravussin, E. & Bogardus, C. (1989). Relationship of genetics, age and physical fitness to daily energy expenditure and fuel utilisation. *American Journal of Clinical Nutrition*, **49**, 968–75.

Sheng, H.P. & Huggins, R.A. (1979). A review of body composition studies with emphasis on total-body water and fat. *American Journal of Clinical Nutrition*, **32**, 630–47.

Siri, W.E. (1956). The gross composition of the human body. *Advances in Biology, Medicine and Physics*, **4**, 239–80.

Snyder, W.S., Cook, M.J., Nasser, E.S., Karhausen, L.R., Howells, G.P. & Tipton, I.H. (1975). *Report of the Task Group on Reference Man*. Oxford: Pergamon.

Wang, Z.M., Pierson, R.N. Jr. & Heymsfield, S.B. (1992). The five level model: a new approach to organising body composition research. *American Journal of Clinical Nutrition*, **56**, 19–28.
Wang, Z.M., Heshka, S., Pierson, R.N. Jr. & Heymsfield, S.B. (1995). Systematic organization of body composition methodology: overview with emphasis on component-based methods. *American Journal of Clinical Nutrition*, in press.

Index

Note: page numbers in *italics* refer to figures and tables.